# A FIRST COURSE IN NUMERICAL ANALYSIS

# VNR NEW MATHEMATICS LIBRARY

under the general editorship of

J. V. ARMITAGE
*Professor of Mathematics*
*University of Nottingham*

N. CURLE
*Professor of Applied Mathematics*
*University of St Andrews*

The aim of this series is to provide a reliable modern coverage of those mainstream topics that form the core of mathematical instruction in universities and comparable institutions. Each book deals concisely with a well-defined key area in pure or applied mathematics or statistics. Many of the volumes are intended not solely for students of mathematics, but also for engineering and science students whose training demands a firm grounding in mathematical methods.

A FIRST COURSE IN

# NUMERICAL ANALYSIS

**M. A. WOLFE**
*Lecturer in Applied Mathematics*
*University of St. Andrews*

VAN NOSTRAND REINHOLD COMPANY
LONDON
NEW YORK  CINCINNATI  TORONTO
MELBOURNE

VAN NOSTRAND REINHOLD COMPANY
25–28 Buckingham Gate, London, SW1E 6LQ

INTERNATIONAL OFFICES
New York    Cincinnati    Toronto    Melbourne

© 1972 M. A. Wolfe
All Rights Reserved. No part of this publication
may be reproduced, stored in a retrieval system,
or transmitted in any form, or by any means,
electronic, mechanical, photocopying, recording,
or otherwise, without the prior permission of
the copyright owner.

Library of Congress Catalog Card No. 72–5929
ISBN: 0 442 09523 6 cloth
      0 442 09524 4 paper

*First Published 1972*

*Printed in Great Britain
by Butler and Tanner Ltd., Frome and London*

# Preface

Numerical analysis is useful to students of those branches of science involving the mathematical analysis of experimental data and its inter-relationships. Numerical analysis is beautiful not the least because it provides a kind of concrete realization of the mathematics which underlies a given problem.

The rapid development of electronic digital computers during the last twenty years has had a profound effect upon the development of numerical analysis. Nowadays virtually all non-trivial numerical calculations are performed on such computers, thereby removing what to some is the drudgery and uncertainty of hand computation and, more significantly, making it possible to use techniques for solving numerical problems hitherto impracticable because of the amount of hand computation involved. As a result of this, textbooks of numerical analysis now contain a preponderance of algorithms suitable for implementation on a computer. Consequently the intending student of numerical analysis should acquire sufficient familiarity with a high-level computer language such as *Fortran* or *Algol* to be able to implement the algorithms to be studied. The purpose of this is twofold. Firstly, human psychology is such that understanding is often enhanced by actually using an algorithm to solve concrete numerical problems. Secondly, an algorithm, albeit satisfying in itself, is useless unless it can be implemented. It should not be forgotten that historically, numerical analysis developed in response to the need to solve numerical problems of increasing complexity arising from the investigation of physical phenomena, as indeed did a great deal of mathematics.

# PREFACE

This book provides, for the undergraduate student of applied mathematics in its widest sense, an elementary introduction to several important topics in numerical analysis.

The mathematical analysis required for a complete understanding of the material in this book will probably have been studied by students of applied mathematics by the end of their first year, and almost certainly by the end of their second. In fact this book is based upon some of the material presented to first- and second-year students of applied mathematics in the University of St Andrews.

The author had two conscious aims in writing this book:
(a) to provide a working knowledge of certain important techniques for the solution of some ubiquitous problems in numerical mathematics;
(b) to give the reader a feeling for the nature of numerical analysis through a few illustrative topics, and thereby to provide a stimulus to explore further the beauty and power of the subject.

The author is grateful to Dr J. A. Glen for pointing out a number of errors and obscurities, and any which still remain are the author's sole responsibility. The author is also grateful to those who typed the manuscript, and in particular to Miss K. P. Dunne whose work was indispensable at this stage.

*St Andrews*
1972
M. A. WOLFE

# Contents

| | |
|---|---|
| Preface | v |

### *Chapter 1. Preliminaries* — 1

| | | |
|---|---|---|
| 1.1 | Introduction | 1 |
| 1.2 | Error in Numerical Calculation | 7 |
| | Tutorial Examples | 17 |

### *Chapter 2. The Interpolating Polynomial* — 20

| | | |
|---|---|---|
| 2.1 | Introduction | 20 |
| 2.2 | Lagrange Interpolation | 22 |
| 2.3 | A Truncation Error Formula for the Interpolating Polynomial | 25 |
| 2.4 | Effect of Rounding Error | 30 |
| 2.5 | Use of Linear Interpolation in Tables | 32 |
| 2.6 | Differences | 33 |
| 2.7 | Newton's Forward Difference Formula | 39 |
| 2.8 | Newton's Backward Difference Formula | 46 |
| | Tutorial Examples | 49 |

### *Chapter 3. Numerical Differentiation and Integration* — 52

| | | |
|---|---|---|
| 3.1 | Introduction | 52 |
| 3.2 | Differentiation using the Interpolating Polynomial | 53 |
| 3.3 | A Truncation Error Formula for Numerical Differentiation | 57 |
| 3.4 | The Taylor Expansion Method for Numerical Differentiation | 59 |
| 3.5 | Rounding Error in Numerical Differentiation | 62 |

| | | |
|---|---|---|
| 3.6 | Numerical Integration using the Interpolating Polynomial | 65 |
| 3.7 | Truncation Error in Interpolatory Quadrature | 71 |
| 3.8 | The Taylor Expansion Method for Numerical Integration | 77 |
| | Tutorial Examples | 79 |

*Chapter 4. Numerical Solution of Equations in One Real Variable*     83

| | | |
|---|---|---|
| 4.1 | Introduction | 83 |
| 4.2 | Existence of a Root | 83 |
| 4.3 | Uniqueness of a Root | 84 |
| 4.4 | The Method of Bisection | 87 |
| 4.5 | The Method of Iteration | 89 |
| 4.6 | Linear Convergence | 94 |
| 4.7 | Aitken's $\Delta^2$ Process | 95 |
| 4.8 | Superlinear Convergence | 97 |
| 4.9 | Newton's Method | 99 |
| 4.10 | The Secant Method | 101 |
| 4.11 | The Method of False Position | 103 |
| | Tutorial Examples | 104 |

*Chapter 5. Numerical Solution of Systems of Linear Algebraic Equations*     108

| | | |
|---|---|---|
| 5.1 | Introduction | 108 |
| 5.2 | Gauss Elimination | 109 |
| 5.3 | Choice of Pivots | 111 |
| 5.4 | Computation of Inverse Matrices | 114 |
| 5.5 | Conditioning | 116 |
| 5.6 | Iterative Improvement of Solutions | 117 |
| 5.7 | The Jacobi Method | 119 |
| 5.8 | The Gauss–Seidel Method | 122 |
| | Tutorial Examples | 125 |

## CONTENTS

*Chapter 6. The Numerical Solution of First-Order Ordinary Differential Equations*    127

   6.1 Introduction    127
   6.2 Difference Equations    128
   6.3 Taylor's Algorithm    130
   6.4 Runge–Kutta Methods    132
   6.5 Analysis of Single-Step Methods    138
   6.6 Adams–Bashforth Methods    141
   6.7 Adams–Moulton Methods    143
       Tutorial Examples    146

ANSWERS TO TUTORIAL EXAMPLES    148

BIBLIOGRAPHY    153

INDEX    155

CHAPTER 1

# Preliminaries

## 1.1 Introduction

The nature of numerical analysis can probably be understood most clearly by considering a number of problems which arise frequently in applications of mathematics to pure and applied science.

### 1.1.1 SOLUTION OF EQUATIONS IN ONE VARIABLE

Given a function $f$ of a single real variable $x$ defined on a given interval $[a, b]$† obtain numerical estimates of the values of $x$ for which $f(x) = 0$. If $x^*$ is a numerical estimate of one such value $x$, find a number $E$ such that

$$|x - x^*| \leqslant E.$$

The number $E$ is called an *error bound* for $x^*$.

*Example 1.1*

Estimate the value of $x$ in $[0, 2]$ for which

$$x = \cos x.$$

### 1.1.2 POLYNOMIAL APPROXIMATION

Given that a function $f$ of a single real variable $x$ has a continuous $(n + 1)$th derivative on a given interval $[a, b]$, and given a set $\{x_k : k = 0, \ldots, n\}$ of values of $x$ in $[a, b]$ together with the set $\{f_k : k = 0, \ldots, n\}$ of corresponding function values [so that $f_k = f(x_k)$ $(k = 0, \ldots, n)$] find a polynomial

† In this book, the notations $(a, b)$ and $[a, b]$ refer to open and closed intervals respectively. That is, $[a, b] = \{x : a \leqslant x \leqslant b\}$, and $(a, b) = \{x : a < x < b\}$.

$p_n$ of degree not greater than $n$ which in some sense approximates $f$ on $[a, b]$. Also, find an error bound $E$ such that

$$|f(x) - p_n(x)| \leqslant E \quad (a \leqslant x \leqslant b).$$

*Example 1.2*

Suppose that the value of $\log(1 + x)$ is given at $x = 0\cdot 0$, $x = 0\cdot 1$, and $x = 0\cdot 2$. Obtain a polynomial $p_2$ of degree 2 such that $p_2(0\cdot 0) = \log(1\cdot 0)$, $p_2(0\cdot 1) = \log(1\cdot 1)$, and $p_2(0\cdot 2) = \log(1\cdot 2)$. Find a number $E$ such that

$$|\log(1 + x) - p_2(x)| \leqslant E \quad (0\cdot 0 \leqslant x \leqslant 0\cdot 2).$$

Such a procedure would make it possible to estimate, say, $\log(1\cdot 15)$ with known accuracy from a knowledge of $\log(1 + x)$ at $x = 0\cdot 0$, $x = 0\cdot 1$, and $x = 0\cdot 2$ only, by evaluating $p_2(0\cdot 15)$.

### 1.1.3 Differentiation

Given the same function $f$ as in problem 2, together with the same information, estimate the value of $f^{(1)}(x_i)$† for some $i$ $(0 \leqslant i \leqslant n)$.

If the estimate obtained is denoted by $f_i^{(1)*}$, find an error bound $E$ such that

$$|f^{(1)}(x_i) - f_i^{(1)*}| \leqslant E.$$

*Example 1.3*

Estimate the value of $f^{(1)}(0\cdot 2)$ from Table 1.1, and obtain an error bound for the resulting estimate, given that $|f^{(3)}(x)| \leqslant 1$ $(0\cdot 0 \leqslant x \leqslant 0\cdot 4)$.

| $x$ | 0·0 | 0·1 | 0·2 | 0·3 | 0·4 |
|---|---|---|---|---|---|
| $f(x)$ | 0·0000 | 0·0998 | 0·1987 | 0·2955 | 0·3894 |

Table 1.1

† In this book, the notation $y^{(k)}(x) = \dfrac{\mathrm{d}^k y(x)}{\mathrm{d}x^k}$ is used.

## 1.1.4 Integration

Given the same function $f$ as in problem 2, together with the same information, estimate the value of $I$ where

$$I = \int_a^b f(x)\, dx. \tag{1.1}$$

If the estimate of $I$ obtained is $I^*$, find an error bound $E$ such that

$$|I - I^*| \leqslant E.$$

*Example 1.4*

(a) Estimate the value of $\int_0^1 \exp(x^2)\, dx$ from a table of values of $\exp(x^2)$ for $x = 0\cdot0(0\cdot1)1\cdot0$ (i.e., for values of $x$ from $0\cdot0$ increasing in steps of $0\cdot1$ up to $1\cdot0$) and give a bound on the error.

It is known that the integral in this problem exists (i.e., that it has a definite value) but it cannot be evaluated analytically (i.e., in terms of a combination of a finite number of the so-called elementary functions of analysis such as $\exp x$, $\log x$, $\sin x$, etc.).

(b) Estimate the value of $\int_0^2 dt/(1+t^4)^2$ from a table of values of $1/(1+t^4)^2$ for $t = 0\cdot0(0\cdot1)2\cdot0$ and obtain an error bound for the estimate.

It is known that

$$\int_0^x \frac{dt}{(1+t^4)^2} = \frac{x}{4(1+x^4)} + \frac{3}{8\sqrt{2}}\left[\tan^{-1}\left\{\frac{x\sqrt{2}}{(1-x^2)}\right\} \right.$$
$$\left. + \tanh^{-1}\left\{\frac{x\sqrt{2}}{(1+x^2)}\right\}\right] \quad (-\infty < x < \infty). \tag{1.2}$$

Such an expression is very tedious to evaluate and indeed there is no need to do so if one of the simple numerical integration formulae for estimating the value of a definite integral is used.

In using integration formulae to estimate the value of $I$ in Eqn (1.1) we evaluate $I^*$ from

$$I^* = \sum_{k=0}^{n} \alpha_k f_k, \qquad (1.3)$$

where $\{\alpha_k: k = 0, \ldots, n\}$ is a set of real numbers which are independent of $f$, and $\{f_k: k = 0, \ldots, n\}$ is a set of function values at the points $\{x_k: k = 0, \ldots, n\}$. Usually, the $x_k$ are equally spaced so that

$$x_k = x_0 + kh \quad (k = 0, 1, \ldots, n).$$

Clearly formula (1.3) is easier to evaluate than formula (1.2) because to evaluate the latter, tables of tangents and hyperbolic tangents are required.

### 1.1.5 Solution of Systems of Linear Algebraic Equations

Given the system of linear algebraic equations

$$\sum_{j=1}^{n} a_{ij} x_j = b_i \quad (i = 1, \ldots, n)$$

where $\{a_{ij}: i, j = 1, \ldots, n\}$ and $\{b_i: i = 1, \ldots, n\}$ are sets of known real numbers, estimate the values of $\{x_j: j = 1, \ldots, n\}$.

*Example 1.5*

Estimate $x_1$, $x_2$, $x_3$, and $x_4$ where
$$\begin{aligned} x_1 + 2x_2 + x_3 + x_4 &= 7, \\ 2x_1 + x_2 + 3x_3 + 2x_4 &= 13, \\ x_1 - 2x_2 + x_3 - 4x_4 &= -7, \\ 3x_1 + 2x_2 - 2x_3 + 3x_4 &= 7. \end{aligned}$$

Such problems occur very frequently in applications of mathematics, especially when $n > 3$ and when the $a_{ij}$ and the $b_i$ are not integers.

### 1.1.6 Solution of First-Order Ordinary Differential Equations

Given that the first-order ordinary differential equation

$$\frac{dy}{dx} = f(x, y), \qquad (1.4a)$$

together with the initial condition

$$y(x_0) = y_0, \tag{1.4b}$$

has a unique solution defined on an interval $[a, b]$ containing $x_0$, estimate the value of $y(x)$ for given values of $x$ in $[a, b]$.

*Example 1.6*

(a) Estimate the value of $y(x)$ for $x = 0.00(0.01)0.20$, where

$$\frac{dy}{dx} = 1 + x^2 + y^2; \quad y(0) = 0. \tag{1.5}$$

(b) Estimate the value of $y(x)$ for $x = 0.0(0.1)1.0$, where

$$\frac{dy}{dx} + 3xy = 1; \quad y(0) = 0. \tag{1.6}$$

It is known that (1.6) is satisfied by

$$y(x) = \exp(-x^3) \int_0^x \exp(t^3) \, dt \tag{1.7}$$

To evaluate $y(x)$ from formula (1.7) it is necessary to evaluate $\exp(-x^3)$ and $\int_0^x \exp(t^3) \, dt$. The integral cannot be evaluated analytically although it is known to exist (i.e., to have a definite value for each value of $x$ in $[0, 1]$). It is more convenient to reject formula (1.7) in favour of one of the numerical methods for the solution of Eqn (1.4) given in Chapter 6.

Problems 1 to 6 are among those which are studied in this book. The study of these problems forms a small but practically important part of the branch of applied mathematics called *numerical analysis*, which is concerned with providing numerical solutions to mathematical problems. A more precise statement of the nature of numerical analysis can be made using the idea of an algorithm.

*Definition 1.1*

An *algorithm* is an unambiguous sequential set of mathematical and logical operations which, when implemented (performed) solve a given mathematical problem.

*Example 1.7*

To illustrate the idea of an algorithm a simple mathematical problem will be proposed and an algorithm will be given for solving it.

Problem: Given that $ax + b = c$, where $a$, $b$, and $c$ are known real numbers, find the value of $x$.

*Algorithm*

(1) If $a = 0$ stop;
(2) if $a \neq 0$ compute $d$, where $d = (c - b)/a$ and go to (3);
(3) set $x = d$ and stop.

Most algorithms are more complicated than this and it is not always obvious that they do, in fact, solve a given mathematical problem. It is not even always obvious that a given algorithm is well defined (i.e., that the operations listed therein are mathematically meaningful). For example, provision must be made in step (1) of the algorithm given in Example 1.7 for the possibility that $a$ has the value zero, in which case $d$ would not be defined.

It is not always obvious that a given mathematical problem has *only* one solution or that it has *even* one solution. Consequently, a theorem should be known with sufficient conditions for the given problem to have one and only one solution (existence and uniqueness theorem). Also, it is usually not obvious that an algorithm which has been proposed for the solution of a given problem will work. Consequently a theorem is needed with sufficient conditions for the algorithm to be effective (convergence theorem).

Finally, an algorithm for providing a numerical solution to a mathematical problem gives, in general, an approximation to the exact numerical solution for reasons which are discussed in Section 1.2. Consequently a theorem is required which gives an error bound on the numerical solution obtained.

The preceding discussion indicates that every mathematical problem which is proposed for numerical solution should be accompanied by an existence and uniqueness theorem, and every algorithm which is proposed for the numerical solution of a mathematical problem must be accompanied by a convergence and error analysis. The complexity of many mathematical problems which arise in practice does not permit these requirements to be satisfied completely, but the ideal remains, and any algorithm for which convergence and error analyses are not given should be regarded as at best empirical.

Using the idea of an algorithm, numerical analysis could be defined as follows.

*Definition 1.2*

*Numerical analysis* is the subject concerned with the construction, analysis, and use of algorithms for the numerical solution of mathematical problems to given degrees of numerical accuracy.

## 1.2 Error in Numerical Calculation

Most algorithms in numerical analysis give approximate results when implemented. Error in numerical calculation arises from essentially four sources: (1) truncation; (2) rounding off; (3) inherent errors in data; (4) mistakes. In this section we consider only (1) and (2) explicitly.

### 1.2.1 Truncation Error

Many of the processes of mathematical analysis involve, at least implicitly, an infinite number of operations. For example, if a function is to be evaluated from a Taylor series, an infinite number of terms would have to be summed to obtain an exact value, a procedure which is clearly impossible in practice. We must truncate the series (i.e., neglect all terms after the first few). In so doing we introduce a *truncation error*.

To illustrate these remarks, suppose that it is required to

evaluate $\exp x$ for any positive value of $x$ using the Taylor series for $\exp x$ about $x = 0$. It is known that

$$\exp x = 1 + x + \frac{x^2}{2!} + \ldots + \frac{x^n}{n!} + \frac{x^{n+1}}{(n+1)!} \exp \xi$$

$$(0 < \xi < x) \quad (1.8)$$

for any value of $x$ in $[0, \infty)$ and for any positive integer $n$. To estimate the value of $\exp x$ we could use

$$f_n(x) = 1 + x + \frac{x^2}{2!} + \ldots + \frac{x^n}{n!}. \quad (1.9)$$

Then we have

$$\exp x = f_n(x) + e_T(x), \quad (1.10)$$

where

$$e_T(x) = \frac{x^{n+1}}{(n+1)!} \exp \xi \quad (0 < \xi < x). \quad (1.11)$$

The truncation error in the estimate $f_n(x)$ of $\exp x$ is $e_T(x)$. From Eqn (1.11)

$$e_T(x) \leqslant \frac{x^{n+1}}{(n+1)!} \exp x. \quad (1.12)$$

For any fixed value of $x$ in $[0, \infty)$, $x^{n+1}/(n+1)! \to 0$ as $n \to \infty$ so $e_T(x) \to 0$ as $n \to \infty$. It would thus appear that the value of $\exp x$ can be calculated with arbitrary precision by making $n$ sufficiently large. In practice the accuracy attainable is limited by rounding error.

## 1.2.2 Rounding Error

Most of the numbers with which we work would contain an infinite number of digits if expressed exactly in decimal form. For example, $\pi = 3 \cdot 1415926535\ 8979323846\ 264\ldots$. Now any computing machine is clearly capable of storing only a finite number of decimal digits. Hence we are forced to work with an approximation to, say $\pi$, which contains only as many digits as the capacity of the computing machine allows. The error introduced in this way is called *rounding error*.

Numerical calculation consists of combining two numbers

by addition, subtraction, multiplication, or division. These operations result in a number which is in general an approximation to the exact result. For example, suppose that a computing machine capable of storing numbers with a maximum of only $n$ decimal digits is used to multiply two such numbers. The resulting product will contain either $2n$ or $2n - 1$ decimal digits, which the machine is incapable of storing, and only the first $n$ digits will be retained.

Essentially two methods exist for representing a number with more than $n$ digits by a number containing exactly $n$ digits. These methods are called *chopping* and *rounding* respectively. The former is used in some electronic digital computers and consists of simply discarding all digits after the $n$th. Rounding is in general the more accurate procedure and a number of methods exist for doing it. Rounding is almost always used in hand calculations and in many electronic digital computers.

Before discussing rounding further we need the ideas contained in the following definitions.

*Definition 1.3*

The number of *significant figures* in a given number is the number of digits which are (assumed to be) correct starting on the left with the first non-zero digit and counting to the right.■

For example, 3·14159 is a representation of $\pi$ which contains six significant figures, 0·00023 contains two significant figures provided that the digits 2 and 3 are correct, and 0·460 contains three significant figures assuming that the digit 0 on the right is correct. A valuable convention in writing decimal numbers is never to write more digits on the right than are known to be correct. For example, if it is not known which digit follows the right-hand zero in 0·450 it would be misleading (and wrong) to write 0·4500.

*Definition 1.4*

The number of *decimal places* in a given decimal number is the number of digits to the right of the decimal point.■

For example, 2·718 is a representation of the base $e$ of natural logarithms which is correct to three decimal places (sometimes written 3D) and four significant figures. The number 1230 may be correct to four significant figures, in which case this could be indicated unambiguously by writing it as $1·230 \times 10^3$. If 1230 is correct to only three significant figures then it could be written as $1·23 \times 10^3$.

We now consider one method for rounding a decimal number to $n$ decimal places.

*Algorithm 1.1* (Rounding to $n$D)

Given a decimal number containing more than $n$ decimal places:

(1) Reject all digits after that in the $n$th decimal place and go to (2);

(2) if the rejected number is less than half a unit in the $n$th decimal place leave the $n$th digit unchanged and go to (5);

(3) if the rejected number is exactly half a unit in the $n$th decimal place, replace the $n$th digit by the nearest even digit and go to (5);

(4) if the rejected number is greater than half a unit in the $n$th decimal place, increase the $n$th digit by unity and go to (5);

(5) stop. ■

*Example 1.8*

(a) On rounding 2·65149 to 3D we obtain 2·651 because the rejected number is 0·49 of a unit in the third decimal place.

(b) On rounding 2·6515 to 3D we obtain 2·652 because the rejected number is exactly 0·5 of a unit in the third decimal place. Also, on rounding 2·05 to 1D we obtain 2·0, but on rounding 2·051 to 1D we obtain 2·1 because the number rejected is 0·51 of a unit in the first decimal place.

(c) On rounding 2·65149 to 4D we obtain 2·6515 because the rejected number is 0·9 of a unit in the fourth decimal place.

The preceding method for rounding has the following features:

(1) A number which is correct to $n$D has a rounding error not greater than $0·5 \times 10^{-n}$.

(2) Because of (3) in Algorithm 1.1 the last digit of a rounded number is more likely to be even than odd and hence is more likely to be divisible by two without further rounding. For example, 2·35 rounded to 1D is 2·4 and 2·4/2 = 1·2 correct to 1D without further rounding.

(3) If each member of a large set of numbers is rounded to $n$D using Algorithm 1.1 then the rounding errors tend to be as often positive as negative. If the members of the set are added, then individual errors are likely to cancel to some extent.

From the preceding discussion it is clear that in general, numerical calculations produce approximate results. Whereas a detailed analysis of truncation error is often feasible, it is seldom so for rounding errors in long calculations. Truncation error depends on the algorithm, while rounding error depends on the computing machine as well as on the methods used to perform the calculation. Nowadays electronic digital computers are used for most non-trivial calculations. Such machines are capable of performing about $10^6$ multiplications per second, so that very large calculations can be done in a few hours. The accumulation of rounding error could therefore be considerable, unless controlled by using a large number significant figures and carefully designed algorithms.

Because the results of numerical calculations are not in general exact we need ways of expressing the accuracy of a numerical result.

*Definition 1.5*

Suppose that a numerical estimate $x^*$ has been obtained for a quantity the exact value of which is $x$.

(1) The *error* $e$ in $x^*$ is defined by

$$e = x - x^*$$

and the *absolute error* $e_A$ in $x^*$ is defined to be $|e|$.

(2) The *relative error* $e_R$ in $x^*$ is defined by

$$e_R = (x - x^*)/x = e/x.$$

(3) The *percentage error* $e_P$ in $x^*$ is defined to be $100|e_R|$. ■

*Example 1.9*

Suppose that $x = \frac{1}{3}$ and $x^* = 0.3333$.
Then $e = 0.\dot{3} - 0.3333 = 0.0000\dot{3} = \frac{1}{3} \times 10^{-4}$,
$e_A = \frac{1}{3} \times 10^{-4}$,
$e_R = (\frac{1}{3} \times 10^{-4})/(\frac{1}{3}) = 10^{-4}$,
$e_P = 100 \times 10^{-4} = 0.01\%$.

Considerable insight into the behaviour of rounding error can be obtained by considering its effect on the sum, difference, product, and quotient of two numbers. Let $x_1^*$ and $x_2^*$ be approximate values of $x_1$ and $x_2$ respectively with errors $e_1$ and $e_2$ so that

$$x_1 = x_1^* + e_1, \quad x_2 = x_2^* + e_2.$$

Then

$$x_1 + x_2 = (x_1^* + x_2^*) + (e_1 + e_2),$$
$$x_1 - x_2 = (x_1^* - x_2^*) + (e_1 - e_2).$$

If $e_S$ and $e_D$ are the errors in the sum and difference of $x_1$ and $x_2$, respectively then

$$e_S = e_1 + e_2, \quad e_D = e_1 - e_2.$$

Hence

$$|e_S| = |e_1 + e_2| \leqslant |e_1| + |e_2|, \tag{1.13}$$

and

$$|e_D| = |e_1 - e_2| \leqslant |e_1| + |e_2|. \tag{1.14}$$

Suppose that $|e_1| \leqslant E_1$ and $|e_2| \leqslant E_2$. Then from Eqns (1.13) and (1.14)

$$|e_S| \leqslant E_1 + E_2, \quad |e_D| \leqslant E_1 + E_2.$$

*Example 1.10*

(a) If $e_1$ and $e_2$ are due solely to rounding error, and $x_1^*$ and $x_2^*$ are representations of $x_1$ and $x_2$ correct to $n$D, then $E_1 = E_2 = 0.5 \times 10^{-n}$. So from Eqn (1.13) and Eqn (1.14) the absolute error in $(x_1^* + x_2^*)$ and $(x_1^* - x_2^*)$ cannot exceed $10^{-n}$.

(b) The absolute rounding errors in 31·4, 17·375, and 2·4956 are not greater than $0.5 \times 10^{-1}$, $0.5 \times 10^{-3}$, and $0.5 \times 10^{-4}$

respectively, so that the sum $31 \cdot 4 + 17 \cdot 375 + 2 \cdot 4956$ could be in error by about $\pm 0 \cdot 05$. Hence the value $51 \cdot 2706$ of the sum could be expressed as $51 \cdot 27 \pm 0 \cdot 05$.

Consider now the error in $x_1^* x_2^*$. We have

$$x_1 x_2 = (x_1^* + e_1)(x_2^* + e_2) = x_1^* x_2^* + e_1 x_2^* + e_2 x_1^* + e_1 e_2.$$

Hence if $e_1 e_2$ is very small compared with the other terms in this sum, then the error $e$ in $x_1^* x_2^*$ is given by

$$e \approx\dagger\ e_1 x_2^* + e_2 x_1^*.$$

Hence the relative error $e_R$ in $x_1^* x_2^*$ is given by

$$e_R \approx e_1/x_1^* + e_2/x_2^*, \tag{1.15}$$

where the exact relative error $(e/x_1 x_2)$ has been approximated by $(e/x_1^* x_2^*)$.

### Example 1.11

The rounding error in $2 \cdot 5$ has a magnitude not greater than $0 \cdot 05$ and the corresponding relative error has a magnitude $0 \cdot 02$. Thus the relative error in $(2 \cdot 5)^2 = 6 \cdot 25$ could be of magnitude $0 \cdot 04$. We say that the number $2 \cdot 5$ is correct to 1 in 50 because $2 \cdot 5/0 \cdot 05 = 50$. By Eqn (1.15) the result $6 \cdot 25$ is correct to about 1 in 25. This is another way of saying that the relative error in $6 \cdot 25$ is about $0 \cdot 04$.

Finally, consider the error in $x_1^*/x_2^*$. We have

$$\frac{x_1}{x_2} = \frac{x_1^* + e_1}{x_2^* + e_2} \approx \frac{x_1^*}{x_2^*} + \frac{e_1}{x_2^*} - \frac{e_2 x_1^*}{x_2^{*2}},$$

on using the binomial theorem to approximate $(1 + e_2/x_2^*)^{-1}$ by $(1 - e_2/x_2^*)$, and neglecting $(e_1 e_2/x_2^{*2})$. Hence the error $e$ and the relative error $e_R$ in $x_1^*/x_2^*$ are given by

$$e \approx (e_1/x_2^*) - (e_2 x_1^*/x_2^{*2}),$$

and

$$e_R \approx (e_1/x_1^*) - (e_2/x_2^*), \tag{1.16}$$

where in Eqn (1.16) the exact relative error has been approxi-

† In this book the symbol $\approx$ is used to denote approximate equality, the precise nature of which will be indicated in the text.

mated by $e/(x_1^*/x_2^*)$. From Eqns (1.15) and (1.16) we can estimate the greatest relative error in $(x_1^* x_2^*)$ and $(x_1^*/x_2^*)$ by using the triangle inequality. We obtain, in both cases, the estimate $(|e_1/x_1^*| + |e_2/x_2^*|)$ for the magnitude of the greatest relative error.

*Example 1.12*

If $x_1^* = x_2^* = 2 \cdot 5$ then $|e_1| \leqslant 0 \cdot 05$ and $|e_2| \leqslant 0 \cdot 05$; so $x_1^*/x_2^*$ could have a relative error of magnitude at most $0 \cdot 04$ on using the estimate of greatest relative error just obtained. Hence we could write $x_1/x_2 = 1 \cdot 00 \pm 0 \cdot 04$, so that $x_1^*/x_2^*$ is correct to 1 in 25.

The preceding discussion of error in the basic operations of arithmetic due to errors in $x_1$ and $x_2$ can be extended to consider the error obtained when evaluating a function of $n$ variables $x_1, \ldots, x_n$ due to errors $e_1, \ldots, e_n$ in these variables. The general idea is illustrated by considering a function $f$ of a single variable $x$ defined on $[a, b]$ with $f^{(1)}(x)$ continuous on $[a, b]$. Then

$$f(x_1) - f(x_2) = f^{(1)}(\alpha)(x_1 - x_2)$$

for all values of $x_1$ and $x_2$ in $[a, b]$, where $\alpha$ lies between $x_1$ and $x_2$.

Now $f^{(1)}(x)$ is bounded on $[a, b]$. Suppose that

$$|f^{(1)}(x)| \leqslant M \quad (a \leqslant x \leqslant b).$$

Then

$$|f(x_1) - f(x_2)| \leqslant M |x_1 - x_2|.$$

Hence if $x^*$ is an estimate of $x$ with error $e$, the corresponding error in $f(x^*)$ is bounded according to

$$|f(x) - f(x^*)| \leqslant M|e|.$$

*Example 1.13*

Suppose that $f(x) = \sin 2x$ is to be evaluated for any value of $x$ in $[0, \pi/2]$ and $x$ is known correct to 1D. Then the error

§1.2]   ERROR IN NUMERICAL CALCULATION   15

in the value of sin $2x^*$ is at most $0\cdot 1$ in magnitude, for $|e| \leqslant 0\cdot 05$, and

$$\left|\frac{\mathrm{d}}{\mathrm{d}x}(\sin 2x)\right| \leqslant 2 \quad (0 \leqslant x \leqslant \pi/2),$$

so

$$|\sin 2x - \sin 2x^*| \leqslant 2 \times 0\cdot 05 = 0\cdot 1 \quad (0 \leqslant x \leqslant \pi/2).$$

Finally in this section we consider a very important source of error in numerical calculations which arises when two very nearly equal numbers are subtracted or when a quotient is formed with a divisor very much smaller than the dividend.

Suppose that $x_1^*$ and $x_2^*$ are very nearly equal, with errors $e_1$ and $e_2$ respectively. Then the relative error $e_R$ in $x_1^* - x_2^*$ is given by

$$e_R \approx (e_1 - e_2)/(x_1^* - x_2^*)$$

where $x_1 - x_2$ has been approximated by $x_1^* - x_2^*$. If $|x_1^* - x_2^*| < |e_1 - e_2|$, then $e_R$ may well be very much greater than either $(e_1/x_1^*)$ or $(e_2/x_2^*)$, so that $x_1^* - x_2^*$ will be correct to fewer significant figures than either $x_1^*$ or $x_2^*$. This effect is called *loss of significance*.

*Example 1.14*

Let $x_1^* = 21\cdot 251$, and $x_2^* = 21\cdot 252$ so that $|e_1| \leqslant 0\cdot 5 \times 10^{-3}$ and $|e_2| \leqslant 0\cdot 5 \times 10^{-3}$ if the errors are due solely to rounding. Then $x_1^* - x_2^* = 0\cdot 001$ and $(e_1/x_1^*) \approx (e_2/x_2^*) \approx 0\cdot 25 \times 10^{-4}$. Also $|e_R| \leqslant (|e_1| + |e_2|)/|x_1^* - x_2^*| = 1$. Hence $|e_R|$ could be very much greater than either $|e_1/x_1^*|$ or $|e_2/x_2^*|$. In fact, $x_1^*$ and $x_2^*$ are correct to *five* significant figures while $x_1^* - x_2^*$ is correct to only *one* significant figure.

Consider now the evaluation of $f(x)$ where

$$f(x) = a/(b^2 - x^2),$$

for a value of $x$ very close to $b$. Let the approximate value $x^*$ with error $\varepsilon$ be used for $x$ where $\varepsilon$ is small compared with $x$. Then the error $e$ in $f(x^*)$ is given by

$$e = f(x) - f(x^*) \approx \frac{d}{dx}\left[\frac{a}{(b^2 - x^{*2})}\right]\varepsilon$$
$$= 2ax^*\varepsilon/(b^2 - x^{*2})^2.$$

Clearly if $x^* \approx b$ and $a > 1$, $e \gg \varepsilon$. The corresponding relative error $e_R$ is given by

$$e_R \approx \frac{2x^*\varepsilon}{(b^2 - x^{*2})} = \left[\frac{2x^{*2}}{(b^2 - x^{*2})}\right]\left(\frac{\varepsilon}{x^*}\right).$$

So $e_R \gg (\varepsilon/x^*)$ when $x^* \approx b$.

*Example 1.15*

Let $a = 1$, $b = 2$, and $x^* = 1\cdot99$ so that $|\varepsilon| \leqslant 0\cdot5 \times 10^{-2}$. Then $(\varepsilon/x^*) \approx 0\cdot25 \times 10^{-2}$ and $e_R \approx 200\,(\varepsilon/x^*)$. Hence although $x^*$ is correct to 1 in 400, $f(x^*)$ is correct to only 1 in 2.

Loss of significance can sometimes be avoided by exploiting some algebraic property of the function to be evaluated.

*Example 1.16*

Suppose that it is required to evaluate $x - a$ for a value of $x$ very nearly equal to $a$. Suppose that $x$ and $a$ are known to be related by

$$x^2 - a^2 = 1.$$

Then

$$x - a = 1/(x + a),$$

and it is better to evaluate $1/(x + a)$ than $(x - a)$ directly. To see why this is so, let

$$f(x) = 1/(x + a),$$

and suppose that $x^*$ is an estimate of $x$ with error $e$. Then

$$|f(x) - f(x^*)| \approx |e/(x^* + a)^2|,$$

and the magnitude of the relative error in $f(x^*)$ is given by

$$\left|\frac{f(x) - f(x^*)}{f(x)^*}\right| \approx \left|\frac{e}{(x^* + a)}\right|$$
$$\approx \left|\frac{e}{2x^*}\right|.$$

However, if $$g(x) = x - a,$$
then
$$\left|\frac{g(x) - g(x^*)}{g(x)}\right| \approx \left|\frac{e}{(x^* - a)}\right|$$
$$= \left|\frac{ex^*}{x^*(x^* - a)}\right|.$$

Clearly if $x^* \approx a$, $x^*/(x^* - a)$ could be very much greater than unity. Hence while the relative error in $f(x^*)$ is actually less than that of $x^*$, the relative error of $g(x^*)$ may be very much greater.

## Tutorial Examples

1. Construct an algorithm for solving the quadratic equation
$$ax^2 + bx + c = 0$$
in which $a$, $b$, and $c$ are real numbers any of which may be zero.

2. Let the polynomial $p_n$ be defined by
$$p_n(x) = \sum_{k=0}^{n} a_k x^k \quad (a_n \neq 0)\,(-\infty < x < \infty),$$
and let $\alpha$ be any real number in $(-\infty, \infty)$. Let the sequence $\{b_i\}$ $(i = n, \ldots, 0)$ be generated recursively from
$$b_n = a_n, \quad b_{n-k} = a_{n-k} + \alpha b_{n-k+1} \quad (k = 1, \ldots, n).$$
Show that $p_n(\alpha) = b_0$. This algorithm is often referred to as *nested multiplication*. Why?

3. By expressing $p_n(x)$ in the form
$$p_n(x) = (x - \alpha)q_{n-1}(x) + b_0$$
where $q_{n-1}$ is a polynomial of degree at most $(n - 1)$, show that
$$q_{n-1}(x) = \sum_{k=1}^{n} b_k x^{k-1}$$

where the $b_k$ are given in Tutorial Example 1.2. The procedure of computing the $b_k$ to obtain $q_{n-1}$ is often called *synthetic division*. Why?

4. Use the result of Tutorial Example 1.3 to construct an algorithm for simultaneously computing $p_n(\alpha)$ and $p_n^{(1)}(\alpha)$. Compare the number of multiplications and additions needed to implement this algorithm with the number needed to evaluate $p_n(\alpha)$ and $p_n^{(1)}(\alpha)$ by forming each term separately in both expressions.

5. If the value of $x$ is given correct to 2D, find the greatest possible number of correct significant figures in the value of $x \exp x$ when $x = 0.25$.

6. Show that one of the roots $\hat{x}$ of the equation
$$ax^2 + bx + c = 0$$
where $a$, $b$, and $c$ are real numbers is given by
$$\hat{x} = -2c/[b + (b^2 - 4ac)^{1/2}].$$
Explain why it is preferable to use this formula rather than the usual one to compute $\hat{x}$ when $b^2 \gg 4ac$, and $b > 0$.

7. It is required to evaluate the following expressions for values of $x$ such that $|x| \gg |\alpha|$ and $\alpha$ is a known constant.
   (a) $(x + \alpha)^{1/2} - x^{1/2}$;
   (b) $[\sin(x + \alpha) - \sin x]/\alpha$;
   (c) $(x + \alpha)^{-1} - x^{-1}$;
   (d) $\log(x + \alpha) - \log x$;
   (e) $\cos(x + \alpha) - \cos x$.

Show how to rearrange these expressions so as to avoid loss of significance.

8. Let $f$ be a function of $n$ variables $x_1, \ldots, x_n$ defined on a domain $a_i \leqslant x_i \leqslant b_i$ $(i = 1, \ldots, n)$ and suppose that $(\partial f/\partial x_i)$ $(i = 1, \ldots, n)$ are all continuous on this domain.

Show that an estimate of the error $e$ which is obtained when computing the value of $f(x_1^*, \ldots, x_n^*)$ in which $x_i^*$ has error $e_i$ ($i = 1, \ldots, n$), is given by

$$e \approx \sum_{i=1}^{n} \frac{\partial f}{\partial x_i}(x_1^*, \ldots, x_n^*) e_i.$$

Apply this result to the functions of $x_1$ and $x_2$ given by

$$f(x_1, x_2) = x_1 \pm x_2;$$
$$f(x_1, x_2) = x_1 x_2;$$
$$f(x_1, x_2) = x_1/x_2.$$

# CHAPTER 2

# The Interpolating Polynomial

## 2.1 Introduction

It is often required to estimate numerical values of a function $f$, its derivative, or its integral. The analytical form of $f$ may be unknown, but a table of values of $f$ may be given. Alternatively, the analytical form of $f$ may be known, but may be very tedious to evaluate, differentiate, or integrate. Frequently $f$ is such that it is impossible to express its integral in a form which is suitable for numerical evaluation. In these situations it is desirable to approximate $f$ in some sense by a function $p$ which is easily evaluated, differentiated, and integrated. Then $f(x)$, $p^{(1)}(x)$, and $\int_a^b p(x)\,dx$ can be regarded as estimates of $f(x)$, $f^{(1)}(x)$, and $\int_a^b f(x)\,dx$ respectively. An obvious choice for $p$ is a polynomial $p_n$ of degree at most $n$ of the form

$$p_n(x) = \sum_{k=0}^{n} a_k x^k. \qquad (2.1)$$

In this chapter we shall consider a method of approximating a continuous function $f$ on a given interval $[a, b]$ by a so-called *interpolating polynomial*. Firstly we need the following definitions.

### Definition 2.1

Let $[a, b]$ be a given closed interval and let $\{x_k : k = 0, \ldots, n\}$ be any $(n + 1)$ distinct points in $[a, b]$. Let $f$ be a function of $x$ defined on $[a, b]$ and let

$$f_k = f(x_k) \quad (k = 0, \ldots, n). \qquad (2.2)$$

If the set $\{f_k: k = 0, \ldots, n\}$ is known than $f$ is said to be *tabulated* at the *tabular points* or *interpolating points* $\{x_k: k = 0, \ldots, n\}$.

*Definition 2.2*

Let $p_n$ be defined by Eqn (2.1), and let the coefficients $\{a_k: k = 0, \ldots, n\}$ be such that

$$p_n(x_k) = f_k \quad (k = 0, \ldots, n). \tag{2.3}$$

Then $p_n$ is called an *interpolating polynomial* for $f$ which interpolates $f$ on the points $\{x_k: k = 0, \ldots, n\}$.

If $p_n$ exists then there is only one such polynomial. This is established by the following theorem.

*THEOREM 2.1*

If there exists a polynomial $p_n$ of degree not greater than $n$ which interpolates a given function $f$ on the set of $(n + 1)$ distinct interpolating points $\{x_k: k = 0, \ldots, n\}$ then this polynomial is unique.

*Proof*

By hypothesis,

$$f(x_k) = f_k = p_n(x_k) \quad (k = 0, \ldots, n). \tag{2.4}$$

Suppose there is another polynomial $q_n$ of degree not greater than $n$ which also interpolates $f$ on $\{x_k: k = 0, \ldots, n\}$. Then

$$f(x_k) = f_k = q_n(x_k) \quad (k = 0, \ldots, n). \tag{2.5}$$

Let $r_n(x)$ be defined by

$$r_n(x) = p_n(x) - q_n(x) \quad (a \leqslant x \leqslant b).$$

Then $r_n$ is a polynomial of degree not greater than $n$, which, by Eqns (2.4) and (2.5) has $(n + 1)$ distinct zeros

$$\{x_k: k = 0, \ldots, n\}.$$

Therefore $r_n(x)$ is identically equal to zero for all values of $x$ in $[a, b]$, whence

$$p_n(x) \equiv q_n(x) \quad (a \leqslant x \leqslant b).$$

Hence the interpolating polynomial $p_n$, if it exists, is unique.

This theorem is important because we shall obtain a number of different expressions for $p_n$ which appear superficially to be distinct polynomials. It is important, however, when considering error analyses to realize that these expressions are in fact different forms of one and the same polynomial $p_n$.

## 2.2 Lagrange Interpolation

In this section the existence of the interpolating polynomial is established by means of a constructive proof (i.e., an existence proof in which the entity the existence of which it is required to establish is actually constructed.)

### THEOREM 2.2

Let $\{x_k : k = 0, \ldots, n\}$ be any $(n + 1)$ distinct points in $[a, b]$ and let $f$ be a function of $x$ defined on $[a, b]$. Then there exists a polynomial $p_n$ of degree not greater than $n$ such that

$$p_n(x_k) = f(x_k) \quad (k = 0, \ldots, n).$$

*Proof*

We proceed by constructing $p_n$ but first we introduce some notation. Let $\{u_k : k = 0, \ldots, n\}$ be any set of $(n + 1)$ numbers. Then the symbols $\prod_{k=0}^{n} u_k$ and $\prod_{\substack{k=0 \\ k \neq j}}^{n} u_k$ are defined by

$$\prod_{k=0}^{n} u_k = u_0 u_1 \ldots u_n, \tag{2.6a}$$

$$\prod_{\substack{k=0 \\ k \neq j}}^{n} u_k = u_0 u_1 \ldots u_{j-1} u_{j+1} \ldots u_n \quad (j = 0, \ldots, n). \tag{2.6b}$$

Also, let $\{L_k(x) : k = 0, \ldots, n\}$ be defined by

$$L_k(x) = \prod_{\substack{j=0 \\ j \neq k}}^{n} \frac{(x - x_j)}{(x_k - x_j)} \quad (a \leqslant x \leqslant b)\, (k = 0, \ldots, n). \tag{2.7}$$

Then

$$L_k(x_k) = 1 \quad (k = 0, \ldots, n), \tag{2.8a}$$

$$L_k(x_j) = 0 \quad (j \neq k;\, j, k = 0, \ldots, n), \tag{2.8b}$$

and $L_k$ is a polynomial in $x$ of degree not greater than $n$. Define the polynomial $p_n$ by

$$p_n(x) = \sum_{k=0}^{n} f_k L_k(x) \quad (a \leqslant x \leqslant b), \qquad (2.9)$$

in which $\{f_k: k = 0, \ldots, n\}$ is defined by Eqn (2.2). Then by Eqns (2.9) and (2.8)

$$p_n(x_j) = f_j \quad (j = 0, \ldots, n),$$

and $p_n$ is a polynomial of degree not greater than $n$. By Theorem 2.1, $p_n$ is therefore the interpolating polynomial which interpolates $f$ on $\{x_k: k = 0, \ldots, n\}$. Hence $p_n$ exists and may be constructed by forming the $L_k(x)$ defined by Eqn (2.7) and using these in Eqn (2.9). ∎

*Example 2.1*

Let us construct the interpolating polynomials $p_1$ and $p_2$ which interpolate $\exp x$ on $\{0, 1\}$ and $\{0, \frac{1}{2}, 1\}$ respectively, given the data in Table 2.1.

| $x$ | 0 | $\frac{1}{2}$ | 1 |
|---|---|---|---|
| $\exp x$ | 1·0000 | 1·6487 | 2·7183 |

TABLE 2.1

Consider firstly $p_1$. We have $x_0 = 0$, $x_1 = 1$, $f_0 = 1 \cdot 0000$, $f_1 = 2 \cdot 7183$. So

$$L_0(x) = \frac{(x - x_1)}{(x_0 - x_1)} = 1 - x,$$

$$L_1(x) = \frac{(x - x_0)}{(x_1 - x_0)} = x,$$

and

$$p_1(x) = 1 \cdot 0000(1 - x) + 2 \cdot 7183(x)$$
$$= 1 \cdot 0000 + 1 \cdot 7183 x \quad (0 \leqslant x \leqslant 1).$$

The procedure of obtaining, from two values of $f$, the interpolating polynomial of degree one is called *linear interpolation*. It is particularly important because, for example, it is used in four-figure tables of functions such as $\exp x$ to interpolate between any two adjacent values of $x$ given in the table.

Consider now $p_2$. We have $x_0 = 0$, $x_1 = \frac{1}{2}$, $x_2 = 1$, $f_0 = 1 \cdot 0000$, $f_1 = 1 \cdot 6487$, and $f_2 = 2 \cdot 7183$. So

$$L_0(x) = \frac{(x - x_1)(x - x_2)}{(x_0 - x_1)(x_0 - x_2)} = 2x^2 - 3x + 1,$$

$$L_1(x) = \frac{(x - x_0)(x - x_2)}{(x_1 - x_0)(x_1 - x_2)} = -4x^2 + 4x,$$

$$L_2(x) = \frac{(x - x_0)(x - x_1)}{(x_2 - x_0)(x_2 - x_1)} = 2x^2 - x,$$

and

$$\begin{aligned}p_2(x) &= 1 \cdot 0000(2x^2 - 3x + 1) + 1 \cdot 6487(-4x^2 + 4x) \\ &\quad + 2 \cdot 7183(2x^2 - x) \\ &= 0 \cdot 8418 x^2 + 0 \cdot 8765 x + 1 \cdot 0000 \quad (0 \leqslant x \leqslant 1).\end{aligned}$$

The polynomials $p_1$ and $p_2$ can clearly be used to approximate $\exp x$ on $[0, 1]$. If the errors $e_1(x)$ and $e_2(x)$ in $p_1(x)$ and $p_2(x)$ respectively, regarded as approximations to $\exp x$ on $[0, 1]$ are defined by

$$e_i(x) = \exp x - p_i(x) \quad (i = 1, 2) \ (0 \leqslant x \leqslant 1),$$

then Table 2.2 gives the values of $e_1$ and $e_2$ on $0 \cdot 0(0 \cdot 2)1 \cdot 0$.

| $x$ | $\exp x$ | $p_1(x)$ | $e_1(x)$ | $p_2(x)$ | $e_2(x)$ |
| --- | --- | --- | --- | --- | --- |
| 0·0 | 1·0000 | 1·0000 | 0·0000 | 1·0000 | 0·0000 |
| 0·2 | 1·2214 | 1·3437 | −0·1223 | 1·2090 | 0·0124 |
| 0·4 | 1·4918 | 1·6873 | −0·1955 | 1·4853 | 0·0065 |
| 0·5 | 1·6487 | 1·8592 | −0·2105 | 1·6487 | 0·0000 |
| 0·6 | 1·8221 | 2·0310 | −0·2089 | 1·8289 | −0·0068 |
| 0·8 | 2·2255 | 2·3746 | −0·1491 | 2·2400 | −0·0145 |
| 1·0 | 2·7183 | 2·7183 | 0·0000 | 2·7183 | 0·0000 |

TABLE 2.2

From Table 2.2 we note that $p_2$ is a much better approximation to exp $x$ than is $p_1$ at the points given. If the values of exp $x$ at points other than the interpolating points used to construct $p_1$ and $p_2$ were not known but were to be estimated using $p_1$ or $p_2$ it would be essential to obtain bounds for $e_1(x)$ and $e_2(x)$ on $[0, 1]$. The errors $e_1(x)$ and $e_2(x)$ arise from two sources. Firstly, exp $x$ is not a polynomial of finite degree so $p_1$ and $p_2$ are essentially different functions from exp $x$ and so their values should not be expected to agree at every point on $[0, 1]$; there is bound to be a truncation error. Secondly, the values of exp $x$ used to construct $p_1$ and $p_2$ are correct to only four decimal places so that each value of exp $x$ is in error by at most $0 \cdot 5 \times 10^{-4}$. As a result of this rounding error, $p_1$ and $p_2$ are not the exact interpolating polynomials. These considerations are taken up in the following sections with greater generality.

## 2.3 A Truncation Error Formula for the Interpolating Polynomial

The interpolating polynomial $p_n$ which interpolates $f$ on $\{x_k: k = 0, \ldots, n\}$ is such that

$$p_n(x_k) = f(x_k) \quad (k = 0, \ldots, n),$$

so that if the truncation error $e(x)$ is defined by

$$e(x) = f(x) - p_n(x),$$

then

$$e(x_k) = 0 \quad (k = 0, \ldots, n). \tag{2.10}$$

Clearly without more information about $f$ nothing can be said about $e(x)$ for $x \neq x_k$ ($k = 0, \ldots, n$). This is illustrated by Fig. 2.1 which shows the graphs of two functions $f_1$ and $f_2$ both of which are interpolated on $\{x_k: k = 0, \ldots, 3\}$ by $p_3$.

If it is known that $f$ has an $(n + 1)$th derivative which is continuous on an interval $[a, b]$ containing the interpolating points $\{x_k: k = 0, \ldots, n\}$ and the point $x$ at which $f(x)$ is to be estimated then $e(x)$ can be expressed in terms of $f^{(n+1)}$. To

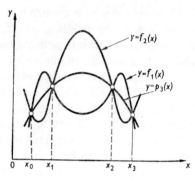

Fig. 2.1

obtain the expression for $e(x)$ certain preliminary results will be needed.

*Lemma 2.1*

Let $\{x_k: k = 0, \ldots, n\}$ be any $(n+1)$ points in $[a, b]$, and let the polynomial $P$ of degree $(n+1)$ be defined by

$$P(x) = \prod_{k=0}^{n} (x - x_k) \quad (a \leqslant x \leqslant b).$$

Then

$$P^{(1)}(x_j) = \prod_{\substack{k=0 \\ k \neq j}}^{n} (x_j - x_k) \tag{2.11}$$

and

$$P^{(n+1)}(x) = (n+1)! \quad (a \leqslant x \leqslant b). \tag{2.12}$$

*Proof*

By hypothesis, $P(x) = (x - x_0) \ldots (x - x_n)$, so by the product rule for differentiation,

$$P^{(1)}(x) = \sum_{k=0}^{n} (x - x_0) \ldots (x - x_{k-1}) \\ \times (x - x_{k+1}) \ldots (x - x_n). \tag{2.13}$$

Now when $k$ is distinct from $j$,

$$(x_j - x_0) \ldots (x_j - x_{k-1})(x_j - x_{k+1}) \ldots (x_j - x_n)$$

is zero because it contains a factor $(x_j - x_j)$. Hence the only term which contributes to the sum in Eqn (2.13) is the term for which $k = j$. Hence Eqn (2.11) is established.

Again by hypothesis,

$$P(x) = x^{n+1} + Q_n(x)$$

where $Q_n$ is a polynomial of degree not greater than $n$ the coefficients of which depend upon $\{x_k: k = 0, \ldots, n\}$. Hence

$$P^{(n+1)}(x) = (n+1)! \quad (a \leqslant x \leqslant b). \blacksquare$$

*Lemma 2.2*

Let $g$ be a function of $x$ defined on an interval $[a, b]$ containing the set $\{s_k: k = 0, \ldots, n\}$ of $(n+1)$ distinct points such that

$$s_0 < s_1 < \ldots < s_n.$$

If

(1) $g^{(n)}$ is continuous on $[s_0, s_n]$;
(2) $g(s_k) = 0$ $(k = 0, \ldots, n)$,

then there exists a number $\xi$ in $(s_0, s_n)$ such that $g^{(n)}(\xi) = 0$.

*Proof*

Applying Rolle's theorem to $[s_{k-1}, s_k]$, we deduce that there exists at least one number $\xi_k^{(1)}$ in $(s_{k-1}, s_k)$ such that $g^{(1)}(\xi_k^{(1)}) = 0$. This statement is true for $k = 1, \ldots, n$. Applying Rolle's theorem to $g^{(1)}(x)$ on $[\xi_{k-1}^{(1)}, \xi_k^{(1)}]$, we deduce that there exists at least one number $\xi_k^{(2)}$ in $(\xi_{k-1}^{(1)}, \xi_k^{(1)})$ such that $g^{(2)}(\xi_k^{(2)}) = 0$. This statement is true for $k = 2, \ldots, n$. By repeated application of Rolle's theorem in this way we deduce that there exists at least one number $\xi_n^{(n)}$ in $(\xi_{n-1}^{(n-1)}, \xi_n^{(n-1)})$ such that $g^{(n)}(\xi_n^{(n)}) = 0$. Clearly, because $s_0 < s_1 < \ldots < s_n$, we have $s_0 < \xi_n^{(n)} < s_n$. Hence taking for $\xi$ the number $\xi_n^{(n)}$, the lemma is proved. $\blacksquare$

Note that Lemma 2.2 is still true when $\{s_k: k = 0, \ldots, n\}$ are not ordered in any special way, for we need only relabel the $s_k$ to make the conditions of the lemma true. The number $\xi$ is then to be found in $(c, d)$ where

$$c = \min_{0 \leqslant k \leqslant n} (s_k) \quad \text{and} \quad d = \max_{0 \leqslant k \leqslant n} (s_k).$$

We are now in a position to obtain a truncation error formula for the interpolating polynomial.

## THEOREM 2.3

Let $\{x_k: k = 0, \ldots, n\}$ be any $(n + 1)$ distinct points in $[a, b]$ which need not be uniformly spaced or in natural order. Let $f$ be a function of $x$ such that $f^{(n+1)}$ is continuous on $[a, b]$ and let $p_n(x)$ be the interpolating polynomial which interpolates $f$ on $\{x_k: k = 0, \ldots, n\}$.

Then corresponding to each value of $x$ in $[a, b]$ there exists a number $\xi$ such that

$$f(x) - p_n(x) = \prod_{k=0}^{n} (x - x_k) \frac{f^{(n+1)}(\xi)}{(n + 1)!}, \qquad (2.14)$$

and

$$\min \{x, x_0, \ldots, x_n\} < \xi < \max \{x, x_0, \ldots, x_n\}.$$

*Proof*

If $x = x_k$ the formula is trivially true. Suppose $x \neq x_k$. Define $q(t)$ and $g(t)$ by

$$q(t) = \prod_{k=0}^{n} (t - x_k) \quad (a \leqslant t \leqslant b),$$

$$g(t) = f(t) - p_n(t) - \frac{q(t)}{q(x)}[f(x) - p_n(x)] \quad (a \leqslant t \leqslant b).$$

Then

$$g(x_k) = 0 \quad (k = 0, \ldots, n),$$

and if $x \neq x_k$ then

$$g(x) = 0.$$

Hence for a fixed value of $x$, $g(t)$ has $(n + 2)$ distinct zeros, namely $x_0, \ldots, x_n$ and $x$, so since all the conditions of Lemma 2.2 are satisfied by $g$, we deduce that there exists a number $\xi$ such that

$$g^{(n+1)}(\xi) = 0,$$

and
$$\min \{x_0, \ldots, x_n, x\} < \xi < \max \{x_0, \ldots, x_n, x\}.$$

Because $q$ is a polynomial of degree at most $(n+1)$ in $t$,
$$q^{(n+1)}(t) = (n+1)!$$
so differentiating $g$ $n+1$ times with respect to $t$ and remembering that $x$ is fixed we obtain
$$g^{(n+1)}(\xi) = f^{(n+1)}(\xi) - (n+1)! \, [f(x) - p_n(x)]/q(x) = 0,$$
whence by rearrangement, Eqn (2.14) follows. ∎

The formula (2.14) cannot usually be used to calculate the truncation error $f(x) - p_n(x)$ because the value of $\xi$ is not known. However, if a bound on $f^{(n+1)}(x)$ is known for $x$ in $[a, b]$, Eqn (2.14) provides a bound on $|f(x) - p_n(x)|$.

*Example* 2.2

Let us obtain bounds on the truncation errors for the polynomials $p_1$ and $p_2$ obtained in Example (2.1).

For $p_1$, $x_0 = 0$, $x_1 = 1$, $n = 1$, $f^{(2)}(x) = \exp x$.

So
$$q(x) = (x - x_0)(x - x_1) = x(x - 1),$$
and
$$|\exp x - p_1(x)| = \left|\frac{x(x-1)}{2} \exp \xi\right| \ (0 \leqslant x \leqslant 1)(0 < \xi < 1).$$

Hence
$$|\exp x - p_1(x)| \leqslant |x(x-1)|(e/2) \quad (0 \leqslant x \leqslant 1).$$

Now $|x(x-1)|$ has its maximum value when $x = \frac{1}{2}$ so we obtain
$$|\exp x - p_1(x)| \leqslant e/8 \quad (0 \leqslant x \leqslant 1).$$

From Table 2.1 the magnitude of the greatest actual error is 0·2105 which is certainly less than $e/8$.

For $p_2$, $x_0 = 0$, $x_1 = \frac{1}{2}$, $x_2 = 1$, $n = 2$, $f^{(3)}(x) = \exp x$, and so
$$q(x) = x(x - \tfrac{1}{2})(x - 1)$$

and
$$|\exp x - p_2(x)| \leq |x(2x-1)(x-1)|(e/12) \quad (0 \leq x \leq 1).$$
If
$$y = x(2x-1)(x-1)$$
then
$$\frac{dy}{dx} = 0 \quad \text{when} \quad x = \frac{(3 \pm \sqrt{3})}{6}.$$
Hence
$$|y| \leq 1/\sqrt{3} \quad (0 \leq x \leq 1),$$
and so
$$|\exp x - p_2(x)| \leq e/(12\sqrt{3}) \quad (0 \leq x \leq 1).$$

## 2.4 Effect of Rounding Error

In Example 2.1 interpolating polynomials $p_1$ and $p_2$ were constructed using data from Table 2.1 in which the function values are correct to only 4D. The rounding errors in the data produce corresponding rounding errors in $p_1$ and $p_2$. To examine this effect for the interpolating polynomial $p_n$ which interpolates $f$ on $\{x_k: k = 0, \ldots, n\}$ in $[a, b]$ suppose that the function value $f_k$ has an error $\varepsilon_k$ ($k = 0, \ldots, n$), so that the exact interpolating polynomial $p_n$, and the polynomial $p_n^*$ obtained from the inexact function values are given by

$$p_n^*(x) = \sum_{k=0}^{n} f_k L_k(x),$$

and

$$p_n(x) = \sum_{k=0}^{n} (f_k + \varepsilon_k) L_k(x).$$

Then

$$|p_n(x) - p_n^*(x)| = |\sum_{k=0}^{n} \varepsilon_k L_k(x)|$$
$$\leq \sum_{k=0}^{n} |\varepsilon_k| \cdot |L_k(x)|.$$

If $\{f_k: k = 0, \ldots, n\}$ is given correct to $m$D then $|\varepsilon_k| \leq 10^{-m}/2$ and so the rounding error $e_R(x)$ is bounded according to

$$|e_R(x)| \leq \frac{10^{-m}}{2} \cdot \sum_{k=0}^{n} |L_k(x)|. \tag{2.15}$$

Consider now the combined effect of truncation and rounding error for $p_n$. Let $e_T(x)$ and $e_R(x)$ be the truncation error given by Eqn (2.14) and the rounding error given by Eqn (2.15) respectively, and let $e(x)$ be the total error. Then

$$|e(x)| = |e_T(x) + e_R(x)| \leq |e_T(x)| + |e_R(x)|.$$

*Example 2.3*

Consider $p_1$ and $p_2$ of Example 2.1. For $p_1$,

$$|e_T(x)| \leq e/8 \quad (0 \leq x \leq 1),$$

and from Eqn (2.15),

$$|e_R(x)| \leq \frac{10^{-4}}{2}(|L_0(x)| + |L_1(x)|)$$

$$= \frac{10^{-4}}{2}(|x| + |1 - x|)$$

$$= \frac{10^{-4}}{2} \quad (0 \leq x \leq 1).$$

Hence the total error $e(x)$ is bounded according to

$$|e(x)| \leq \frac{e}{8} + \frac{10^{-4}}{2} \quad (0 \leq x \leq 1).$$

For $p_2$,

$$|e_T(x)| \leq e/(12\sqrt{3}) \quad (0 \leq x \leq 1),$$

and

$$|e_R(x)| \leq \frac{10^{-4}}{2}(|L_0(x)| + |L_1(x)| + |L_2(x)|)$$

$$= \frac{10^{-4}}{2}(|2x^2 - 3x + 1| + |4x(x-1)| + |x(2x-1)|)$$

$$\leq \frac{3 \times 10^{-4}}{2} \quad (0 \leq x \leq 1).$$

Hence the total error $e(x)$ is bounded according to

$$|e(x)| \leqslant \frac{e}{12\sqrt{3}} + \frac{3 \times 10^{-4}}{2} \quad (0 \leqslant x \leqslant 1).$$

## 2.5 Use of Linear Interpolation in Tables

Since tables of values of the elementary functions of analysis can be given for discrete values of the argument $x$ only, it is necessary to interpolate between tabulated function values. Clearly, if linear interpolation is to be used to estimate a function value $f(x)$ from the values $f(x_k)$ and $f(x_{k+1})$, where $x_k < x < x_{k+1}$ then $x_{k+1}$ must be sufficiently close to $x_k$ to ensure that the truncation error incurred by linear interpolation is not greater than $10^{-m}/2$ where $m$ is the number of decimal places to which $f$ is tabulated.

Suppose that $f(x)$ is tabulated on $[a, b]$ at $n+1$ uniformly spaced tabular points $\{x_k: k = 0, \ldots, n\}$ with spacing $h$, and suppose that

$$|f^{(2)}(x)| \leqslant M_k \quad (x_k \leqslant x \leqslant x_{k+1}) \ (k = 0, \ldots, n-1). \quad (2.16)$$

If the value of $f(x)$ is to be estimated by linear interpolation, where $x_k < x < x_{k+1}$ then from Eqns (2.14) and (2.16)

$$|f(x) - p_1(x)| \leqslant |(x - x_k)(x - x_{k+1})|(M_k/2).$$

By differentiating the right-hand side of this inequality

$$\max_{x_k \leqslant x \leqslant x_{k+1}} |f(x) - p_1(x)| \leqslant (x_{k+1} - x_k)^2 (M_k/8)$$
$$= h^2 M_k/8. \quad (2.17)$$

Clearly, if

$$|f^2(x)| \leqslant M \quad (a \leqslant x \leqslant b)$$

then the magnitude of the truncation error $e_T$ incurred by using linear interpolation between *any* two consecutive entries of the table is bounded according to

$$|e_T| \leqslant Mh^2/8. \quad (2.18)$$

If $f$ is tabulated to $m$D then $h$ must therefore satisfy the inequality

$$\frac{Mh^2}{8} \leqslant \frac{10^{-m}}{2} \qquad (2.19)$$

in order that the truncation error should not exceed the rounding error.

*Example 2.4*

If exp $x$ is to be tabulated on [0, 1] to 2D then we need a spacing $h$ between tabular points such that

$$h^2 \leqslant \frac{4 \times 10^{-2}}{e} < 2 \times 10^{-2}$$

because $M = e$. Hence $h = 0\cdot1$ would ensure that the truncation error incurred by using linear interpolation between any two tabular points is not greater than the rounding error in the tabulated values.

## 2.6 Differences

For all of the applications of the theory of interpolation in this book the interpolating points $x_k$ have uniform spacing $h$, so that

$$x_k = x_0 + kh \quad (k = 0, \ldots, n).$$

For this reason it is desirable to obtain forms of the interpolating polynomial for uniformly spaced interpolating points, in terms of the so-called forward and backward differences of the function $f$ which is to be interpolated.

*Definition 2.3*

Let $f$ be defined on $[a, b]$ and let $x$ and $x + h$ be in $[a, b]$, where $h > 0$.
Then the *forward difference operator* $\Delta$ is defined by

$$\Delta f(x) = f(x + h) - f(x). \qquad (2.20)$$

Repeated application of the forward difference operator $\Delta$ is defined recursively as follows.

## Definition 2.4

Let $f$ be defined on $[a, b]$ and let $x$ and $x + mh$ be in $[a, b]$ where $m$ is a positive integer and $h$ is positive. Then the operator $\Delta^k$ is defined by

$$\Delta^0 f(x) = f(x), \tag{2.21a}$$

$$\Delta^k f(x) = \Delta(\Delta^{k-1} f(x)) \quad (k = 1, 2, \ldots, m). \blacksquare \tag{2.21b}$$

If $f$ is tabulated on $\{x_k : k = 0, \ldots, n\}$ then $\Delta^k f(x_r)$ is called the *kth order forward difference* of $f$ at $x_r$ and is usually written $\Delta^k f_r$. Clearly $\Delta^k f_r$ can be expressed in terms of the tabular values of $f$.

## Example 2.5

(1) $\Delta f_r = f_{r+1} - f_r$;

(2) $\Delta^2 f_r = \Delta(\Delta f_r) = (f_{r+2} - f_{r+1}) - (f_{r+1} - f_r)$
$= f_{r+2} - 2f_{r+1} + f_r$;

(3) $\Delta^3 f_r = \Delta(\Delta^2 f_r)$
$= (f_{r+3} - 2f_{r+2} + f_{r+1}) - (f_{r+2} - 2f_{r+1} + f_r)$
$= f_{r+3} - 3f_{r+2} + 3f_{r+1} - f_r$.

In general we have the following theorem.

## THEOREM 2.4

With the notation previously defined,

$$\Delta^m f_r = \sum_{k=0}^{m} \binom{m}{k} (-1)^k f_{m+r-k}, \tag{2.22}$$

where $\binom{m}{k} = \dfrac{m!}{k!\,(m-k)!}$. $\blacksquare$

To prove this theorem we need the following lemma.

## Lemma 2.3

If $k$ and $p$ are integers then

$$\binom{k}{p} + \binom{k}{p-1} = \binom{k+1}{p} \quad (k > 0, p > 0).$$

*Proof*

$$\binom{k}{p} + \binom{k}{p-1} = \frac{k!}{p!(k-p)!} + \frac{k!}{(p-1)!(k-p+1)!}$$

$$= \frac{k!}{(p-1)!(k-p)!}\left[\frac{k+1}{p(k-p+1)}\right]$$

$$= \frac{(k+1)!}{p!(k-p+1)!}$$

$$= \binom{k+1}{p}. \blacksquare$$

*Proof of Theorem 2.4*

The relation (2.22) is obviously true for $m = 1$. Assume it true for $m = 1, \ldots, p$. Then

$$\Delta^{p+1}f_r = \Delta(\Delta^p f_r) = \sum_{k=0}^{p}(-1)^k \binom{p}{k}f_{p+r-k+1}$$

$$- \sum_{k=0}^{p}(-1)^k \binom{p}{k}f_{p+r-k}$$

$$= \sum_{k=0}^{p}(-1)^k \binom{p}{k}f_{p+r-k+1}$$

$$+ \sum_{k=1}^{p+1}(-1)^k \binom{p}{k-1}f_{p+r-k+1}$$

$$= f_{p+r+1} + \sum_{k=1}^{p}(-1)^k \binom{p+1}{k}f_{p+1+r-k}$$

$$+ (-1)^{p+1}f_r$$

$$= \sum_{p=0}^{p+1}(-1)^k \binom{p+1}{k}f_{p+1+r-k},$$

on using Lemma 2.3.

Hence Eqn (2.22) is true for $m = p + 1$ if it is true for $m = 1, \ldots, p$. Hence by induction Eqn (2.22) is true for all values of $m$. $\blacksquare$

Although Eqn (2.22) is useful for theoretical purposes it is not used to obtain forward differences numerically. The forward difference table is constructed using Eqn. (2.21) in the form

$$\Delta^{m+1}f_k = \Delta^m f_{k+1} - \Delta^m f_k. \qquad (2.23)$$

| $x_k$ | $f_k$ | $\Delta f_k$ | $\Delta^2 f_k$ | $\Delta^3 f_k$ | $\Delta^4 f_k$ |
|---|---|---|---|---|---|
| ⋮ | | | | | |
| $x_{-2}$ | $f_{-2}$ | | | | |
| | | $\Delta f_{-2}$ | | | |
| $x_{-1}$ | $f_{-1}$ | | $\Delta^2 f_{-2}$ | | |
| | | $\Delta f_{-1}$ | | $\Delta^3 f_{-2}$ | |
| $x_0$ | $f_0$ | | $\Delta^2 f_{-1}$ | | $\Delta^4 f_{-2}$ |
| | | $\Delta f_0$ | | $\Delta^3 f_{-1}$ | |
| $x_1$ | $f_1$ | | $\Delta^2 f_0$ | | |
| | | $\Delta f_1$ | | | |
| $x_2$ | $f_2$ | | | | |
| ⋮ | | | | | |

TABLE 2.3

In Table 2.3 we calculate, in particular, $\Delta f_2$, $\Delta^2 f_{-2}$, $\Delta^3 f_{-2}$, and $\Delta^4 f_{-2}$ from

$$\Delta^{m+1}f_{-2} = \Delta^m f_{-1} - \Delta^m f_{-2} \quad (m = 0, \ldots, 3). \qquad (2.24)$$

*Example* 2.6

A forward difference table for the function $f$ defined by

$$f(x) = x^3 \quad (-3 \leqslant x \leqslant 3)$$

is shown in Table 2.4.

| $x_k$ | $f_k$ | $\Delta f_k$ | $\Delta^2 f_k$ | $\Delta^3 f_k$ | $\Delta^4 f_k$ | $\Delta^5 f_k$ | $\Delta^6 f_k$ |
|---|---|---|---|---|---|---|---|
| $-3$ | $-27$ | | | | | | |
| | | 19 | | | | | |
| $-2$ | $-8$ | | $-12$ | | | | |
| | | 7 | | 6 | | | |
| $-1$ | $-1$ | | $-6$ | | 0 | | |
| | | 1 | | 6 | | 0 | |
| 0 | 0 | | 0 | | 0 | | 0 |
| | | 1 | | 6 | | 0 | |
| 1 | 1 | | 6 | | 0 | | |
| | | 7 | | 6 | | | |
| 2 | 8 | | 12 | | | | |
| | | 19 | | | | | |
| 3 | 27 | | | | | | |

TABLE 2.4

It is convenient for subsequent applications to introduce the *backward difference operator* $\nabla$ defined as follows.

*Definition 2.5*

Let $f$ be defined on $[a, b]$ and let $x$ and $x - h$ be in $[a, b]$ where $h > 0$.

Then the *backward difference operator* $\nabla$ is defined by
$$\nabla f(x) = f(x) - f(x - h). \qquad (2.25)$$

*Definition 2.6*

Let $f$ be defined on $[a, b]$ and let $x$ and $x - mh$ be in $[a,b]$ where $h > 0$ and $m$ is a positive integer. Then the operator $\nabla^k$ is defined by
$$\nabla^0 f(x) = f(x), \qquad (2.26a)$$
$$\nabla^k f(x) = \nabla(\nabla^{k-1} f(x)) \quad (k = 1, \ldots, m). \qquad (2.26b)$$

If $f$ is tabulated on $\{x_k : k = 0, \ldots, n\}$ then $\nabla^k f(x_r)$ is called the *kth order backward difference* of $f$ at $x_r$, and is written $\nabla^k f_r$. Clearly $\nabla^k f_r$ can be expressed in terms of the tabular values of $f$.

*Example 2.7*

(1) $\nabla f_r = f_r - f_{r-1}$;

(2) $\nabla^2 f_r = \nabla(\nabla f_r) = (f_r - f_{r-1}) - (f_{r-1} - f_{r-2})$
$= f_r - 2f_{r-1} + f_{r-2}$;

(3) $\nabla^3 f_r = \nabla(\nabla^2 f_r) = (f_r - 2f_{r-1} + f_{r-2})$
$- (f_{r-1} - 2f_{r-2} + f_{r-3})$
$= f_r - 3f_{r-1} + 3f_{r-2} - f_{r-3}$.

In general we have the following theorem, which is proved by induction in the same way as Theorem 2.4.

*THEOREM 2.5*

With the notation previously defined,
$$\nabla^m f_r = \sum_{k=0}^{m} \binom{m}{k} (-1)^k f_{r-k}. \qquad (2.27)$$

The backward difference table of $f$ is constructed by using Eqns (2.26) in the form

$$\nabla^{m+1}f_k = \nabla^m f_k - \nabla^m f_{k-1}, \qquad (2.28)$$

and is shown for $k = -2, -1, 0, 1, 2$ in Table 2.5.

| $x_k$ | $f_k$ | $\nabla f_k$ | $\nabla^2 f_k$ | $\nabla^3 f_k$ | $\nabla^4 f_k$ |
|---|---|---|---|---|---|
| $x_{-2}$ | $f_{-2}$ | | | | |
| | | $\nabla f_{-1}$ | | | |
| $x_{-1}$ | $f_{-1}$ | | $\nabla^2 f_0$ | | |
| | | $\nabla f_0$ | | $\nabla^3 f_1$ | |
| $x_0$ | $f_0$ | | $\nabla^2 f_1$ | | $\nabla^4 f_2$ |
| | | $\nabla f_1$ | | $\nabla^3 f_2$ | |
| $x_1$ | $f_1$ | | $\nabla^2 f_2$ | | |
| | | $\nabla f_2$ | | | |
| $x_2$ | $f_2$ | | | | |

TABLE 2.5

Clearly the entries in Tables 2.3 and 2.5 are identical so that, for example,

$$\nabla^4 f_2 = \Delta^4 f_{-2}.$$

In general we have the following theorem.

*THEOREM 2.6*

With the notation previously defined,

(1) $\qquad\qquad \nabla^m f_k = \Delta^m f_{k-m}, \qquad (2.29a)$

(2) $\qquad\qquad \Delta^m f_k = \nabla^m f_{k+m}. \qquad (2.29b)$

*Proof*

An inductive proof is easy and is therefore left to the reader. ∎

We end this section by considering the differences of a polynomial. We note that in Table 2.4 the third-order differences are constant while all differences of higher order are zero. This is a special case of the following theorem.

*THEOREM 2.7*

Let $p_n$ be a polynomial in $x$ of degree $n$ defined by

$$p_n(x) = \sum_{k=0}^{n} a_k x^k \quad (a_n \neq 0).$$

Then for all values of $x$,

(1) $\qquad \Delta^n p_n(x) = a_n . h^n . n!,$ \hfill (2.30a)

(2) $\qquad \Delta^m p_n(x) = 0 \quad (m > n).$ \hfill (2.30b)

*Proof*

By the Binomial Theorem

$$\Delta x^k = (x+h)^k - x^k = kx^{k-1}h + \sum_{j=2}^{k} \binom{k}{j} x^{k-j} h^j$$
$$= kx^{k-1}h + q_{k-2}(x) \ (k = 1, 2, 3, \ldots),$$

where $q_{k-2}$ is a polynomial of degree $k - 2$, defined by

$$q_{-1}(x) = 0$$
$$q_k(x) = \sum_{j=2}^{k} \binom{k}{j} x^{k-j} h^j \quad (k = 2, 3, \ldots).$$

Hence by Eqn (2.21b), used repeatedly,

$$\Delta^n x^k = k(k-1) \ldots (k-n+1) x^{k-n} h^n + q_{k-n-1}(x),$$

whence

$$\Delta^k x^k = k! \, h^k,$$

and

$$\Delta^n x^k = 0 \quad (n > k).$$

Then

$$\Delta^n p_n(x) = \sum_{k=0}^{n} a_n \Delta^n x^k = a_n n! \, h^n,$$

and

$$\Delta^m p_n(x) = 0 \quad (m > n). \blacksquare$$

## 2.7 Newton's Forward Difference Formula

The Lagrange form of the interpolating polynomial has the disadvantage that if an additional tabular value becomes available then the polynomials $L_k(x)$ must be recalculated to incorporate the new tabular point. As we have seen, difference tables of a tabulated function are easily constructed. It will be shown that a form of the interpolating polynomial exists which is expressible in terms of the forward differences and which is very easily modified to take into account additional tabular values of the function to be interpolated. The form of the interpolating polynomial referred to is called *Newton's forward difference formula*. To establish this formula we require the following lemma:

*Lemma 2.4*

Let a function $f$ be tabulated on the set $\{x_k: k = 0, \ldots, n\}$, with tabular values $\{f_k: k = 0, \ldots, n\}$. Then

$$f_k = \sum_{m=0}^{k} \binom{k}{m} \Delta^m f_0 \quad (k = 0, \ldots, n). \tag{2.31}$$

*Proof*

Obviously Eqn (2.31) is true for $k = 0$. Assume it true for $k = 0, \ldots, p - 1$ ($p - 1 < n$). Then

$$\begin{aligned}
f_p = f_{p-1} + \Delta f_{p-1} &= \sum_{m=0}^{p-1} \binom{p-1}{m}[\Delta^m f_0 + \Delta^{m+1} f_0] \\
&= \sum_{m=0}^{p-1} \binom{p-1}{m} \Delta^m f_0 + \sum_{m=1}^{p} \binom{p-1}{m-1} \Delta^m f_0 \\
&= f_0 + \sum_{m=1}^{p-1} \left[\binom{p-1}{m} + \binom{p-1}{m-1}\right] \Delta^m f_0 + \Delta^p f_0 \\
&= f_0 + \sum_{m=1}^{p-1} \binom{p}{m} \Delta^m f_0 + \Delta^p f_0 \\
&\qquad\qquad\qquad\qquad\qquad\text{[by Lemma 2.3]} \\
&= \sum_{m=0}^{p} \binom{p}{m} \Delta^m f_0.
\end{aligned}$$

Hence Eqn (2.31) is true for $k = p$ if it is true for $k = 0, \ldots, p - 1$, so by induction it is true for all $k$ for which $f_k$ is defined. ∎

*Example 2.8*

We have
$$f_1 = f_0 + f_1 - f_0 = f_0 + \Delta f_0,$$
and
$$f_2 = f_0 + 2f_1 - 2f_0 + f_2 - 2f_1 + f_0 = f_0 + 2\Delta f_0 + \Delta^2 f_0.$$
Setting $k = 1$ and $k = 2$ in Eqn (2.31) we obtain the same results.

Using Lemma 2.4 we can now obtain Newton's forward difference formula, but first we define the *Binomial Function*.

*Definition 2.7*

Let $s$ be any real number, not necessarily an integer. Then the *binomial function* $\binom{s}{r}$ is defined by

$$\binom{s}{r} = \begin{cases} 1 & (r = 0) \\ s(s-1)\ldots(s-r+1)/r! & (r \text{ is any positive integer}) \\ 0 & (r \text{ is any negative integer}). \end{cases}$$

*THEOREM 2.8*

Let $f$ be defined on $[a, b]$ and tabulated on $\{x_k : k = 0, \ldots, n\}$, where $x_0$ is in $[a, b]$,
$$x_k = x_0 + kh \quad (h > 0) \quad (k = 0, \ldots, n),$$
and $h$ is such that $x_k$ is in $[a, b]$ ($k = 0, \ldots, n$). Also, let
$$x = x_0 + sh,$$
where $s$ is a real number such that $x$ is in $[a, b]$.

Then the polynomial $p_n$ which interpolates $f$ on $\{x_k: k = 0, \ldots, n\}$ is given by
$$p_n(x) = \sum_{k=0}^{n} \binom{s}{k} \Delta^k f_0. \tag{2.32}$$

*Proof*

Let $P_k(s)$ be defined by
$$P_k(s) = p_k(x) \quad (k = 0, \ldots, n) \ (a \leqslant x \leqslant b)$$

where $p_k(x)$ interpolates $f$ on $\{x_0, \ldots, x_k\}$. Then if Eqn (2.32) is true, we have

$$P_0(s) = f_0, \qquad (2.33a)$$

$$P_k(s) = P_{k-1}(s) + \binom{s}{k}\Delta^k f_0 \quad (k = 1, \ldots, n). \qquad (2.33b)$$

We will establish Eqns (2.33) by induction. From Eqn (2.9) we have

$$\begin{aligned}P_1(s) = p_1(x) &= f_0 \frac{(x - x_1)}{(x_0 - x_1)} + f_1 \frac{(x - x_0)}{(x_1 - x_0)} \\ &= f_0(1 - s) + f_1 s \\ &= f_0 + s(f_1 - f_0) = f_0 + s\Delta f_0.\end{aligned}$$

Hence Eqn (2.33b) is true for $k = 1$. Assume it true for $k = 1, \ldots, q - 1$ ($q < n$). Define $G_q(s)$ by

$$G_q(s) = P_{q-1}(s) + \binom{s}{q}\Delta^q f_0.$$

Then by definition of the interpolating polynomial,

$$P_{q-1}(r) = f_r \quad (r = 0, \ldots, q - 1),$$

and since by Definition 2.7,

$$\binom{r}{q} = 0 \quad (r = 0, \ldots, q - 1),$$

we have

$$G_q(r) = f_r \quad (r = 1, \ldots, q - 1).$$

Also

$$G_q(q) = P_{q-1}(q) + \Delta^q f_0.$$

Now by hypothesis, using Eqn (2.32) which we assume to be true for $k = 0, \ldots, q - 1$,

$$P_{q-1}(q) = \sum_{k=0}^{q-1} \binom{q}{k}\Delta^k f_0,$$

so

$$G_q(q) = \sum_{k=0}^{q} \binom{q}{k}\Delta^k f_0 = f_q, \quad \text{by Lemma 2.4.}$$

Hence we have established that

$$G_q(r) = f_r \quad (r = 0, \ldots, q).$$

Furthermore it is clear from the definition of $G_q(s)$ that $G_q(s)$ is a polynomial in $s$ of degree not greater than $q$, and is therefore also a polynomial $g_q$ in $x$ of degree not greater than $q$ which interpolates $f$ on $\{x_r: r = 0, \ldots, q\}$. Hence by Theorem 2.1 we have $G_q(s) = P_q(s)$ and so Eqn (2.33b) is true for $k = q$ if it is true for $k = 1, \ldots, q-1$, and so by induction is true for all $k(0 \leqslant k \leqslant n)$. The truth of Eqn (2.33a) is obvious. ∎

From Theorem 2.8 we have the following algorithm for constructing Newton's forward difference form of the interpolating polynomial $p_n(x)$ or $P_n(s)$.

*Algorithm 2.1*

1. Set $s = (x - x_0)/h$, and $P_0(s) = f_0$.
2. For $k = 1, \ldots, n$ generate the $P_k(s)$ from

$$P_k(s) = P_{k-1}(s) + \binom{s}{k}\Delta^k f_0. \ \blacksquare$$

*Example 2.9*

We will use Newton's formula to estimate the value of $f(0\cdot 1)$ for the function $f$ tabulated at $x = 0\cdot 0(0\cdot 2)0\cdot 6$ in Table 2.6. To use Newton's forward difference formula we need as many forward differences of $f_0$ as possible so we take $x_0$ to be $0\cdot 0$ as shown.

Note that in Table 2.6 only the last significant figures of the differences are recorded as is usual when differences are used in hand calculations. The fact that the differences $\Delta^k f$ decrease as $k$ increases is an indication that no mistakes have been made in recording the tabular values $f_k$ [see Tutorial Example 2.8], and that the interval at which $f$ is tabulated is sufficiently small. The fact that the differences become more nearly constant as $k$ increases indicates that $f$ is increasingly better approximated by the interpolating polynomial of degree $k$ as $k$ increases. [See Tutorial Example 2.7.]

44                THE INTERPOLATING POLYNOMIAL                [Ch. 2

| k | $x_k$ | $f_k$  | $\Delta f_k$ | $\Delta^2 f_k$ | $\Delta^3 f_k$ | $\Delta^4 f_k$ |
|---|-------|--------|--------------|----------------|----------------|----------------|
| 0 | 0·0   | 0·5000 |              |                |                |                |
|   |       |        | −454         |                |                |                |
| 1 | 0·2   | 0·4545 |              | 76             |                |                |
|   |       |        | −378         |                | −19            |                |
| 2 | 0·4   | 0·4167 |              | 57             |                | 8              |
|   |       |        | −321         |                | −11            |                |
| 3 | 0·6   | 0·3846 |              | 46             |                |                |
|   |       |        | −275         |                |                |                |
| 4 | 0·8   | 0·3571 |              |                |                |                |

TABLE 2.6

Now $$s = \frac{(x - x_0)}{h} = \frac{(0\cdot 1 - 0\cdot 0)}{0\cdot 2} = \frac{1}{2},$$

so

$$P_0(s) = f_0 = 0\cdot 5000,$$

$$P_1(s) = P_0(s) + \binom{s}{1}\Delta f_0 = 0\cdot 5000 - \frac{0\cdot 0454}{2} = 0\cdot 4773,$$

$$P_2(s) = P_1(s) + \binom{s}{2}\Delta^2 f_0 = 0\cdot 4773 - \frac{0\cdot 0076}{8} = 0\cdot 4763,$$

$$P_3(s) = P_2(s) + \binom{s}{3}\Delta^3 f_0 = 0\cdot 4763 - \frac{0\cdot 0019}{16} = 0\cdot 4762$$

Hence the estimated value of $f(0\cdot 1)$ is $0\cdot 4762$ using all of the available differences. Note that what has been done in this calculation is to compute, recursively, the values of the interpolating polynomials which interpolate $f$ on $\{x_0\}$, $\{x_0, x_1\}$, $\{x_0, x_1, x_2\}$, and $\{x_0, x_1, x_2, x_3\}$ at $x = 0\cdot 1$.

If now another tabular value at $x_4$ becomes available, it can be incorporated easily into the calculation to obtain an improved estimate of $f(0\cdot 1)$ as follows.

$$P_4(s) = P_3(s) + \binom{s}{4}\Delta^4 f_0 = 0\cdot 4762 - \frac{5 \times 0\cdot 0008}{128} = 0\cdot 4762$$

Note that the additional point makes no difference to the fourth decimal place in this case.

## §2.7] NEWTON'S FORWARD DIFFERENCE FORMULA

Newton's forward difference formula is used for interpolating near the beginning of a table for the following reasons:

(a) as many forward differences of $f_0$ as possible are available;

(b) the magnitude of $s$ is relatively small so that provided $\Delta^k f_0$ decreases sufficiently rapidly as $k$ increases, successive terms $\binom{s}{k}\Delta^k f_0$ of Newton's formula decrease rapidly as $k$ increases.

Newton's forward difference formula requires the tabular points to be uniformly spaced. Its principal non-computational importance lies in its use for obtaining approximate differentiation and integration formulae, as is explained in Chapter 3.

Since Newton's forward difference formula is just another form of the interpolating polynomial, Eqn (2.14) can be used to obtain the truncation error $e_n(x)$. Since

$$x = x_0 + sh, \quad x_k = x_0 + kh \ (k = 0, \ldots, n), \quad p_n(x) = P_n(s),$$

we have, on using these in Eqn (2.14), and using Definition 2.7

$$e_n(x) = f(x) - p_n(x) = f(x) - P_n(s) = \binom{s}{n+1} h^{n+1} f^{(n+1)}(\xi) \tag{2.34}$$

where

$$\min\{x_0, x\} < \xi < \max\{x_n, x\}.$$

*Example 2.10*

The function tabulated in Table 2.6 is the function $f$ defined by

$$f(x) = \frac{1}{(x+2)} \quad (0 \leqslant x < \infty).$$

For $n = 3$,

$$f(x) - P_3(s) = \binom{s}{4} h^4 f^{(4)}(\xi).$$

For $x_0 < x < x_3$, we have $x_0 < \xi < x_3$ and so

$$|f^{(4)}(x)| = \left|\frac{24}{(x+2)^5}\right| \leq \frac{3}{4} \quad (x_0 \leq x \leq x_3).$$

So

$$|e_3(x)| = |f(x) - P_3(s)| \leq \binom{s}{4}\frac{3h^4}{4} \quad (0 \leq s \leq 3).$$

When $x = 0.1$, $s = \frac{1}{2}$ so we have with $h = 0.2$

$$|e_3(0.1)| < 0.00005.$$

Hence the truncation error in the estimate 0.4762 of $f(0.1)$ is less than half a unit in the fourth decimal place. This is easily verified by calculating $f(0.1)$ directly whence

$$f(0.1) = 0.476190\ldots$$

## 2.8 Newton's Backward Difference Formula

For interpolation near the end of a table Newton's forward difference formula cannot be used because sufficient forward differences are not available if we take $x_0$ near the end of the table. If we take $x_0$ at the beginning of the table so that more forward differences are available then $s$ is so large that successive terms of Newton's formula will not decrease rapidly enough to ensure a sufficiently accurate estimate. To overcome these difficulties Newton's backward difference formula can be used. To establish this result we need the following lemma.

*Lemma 2.5*

With the notation previously defined,

$$f_{n-m} = \sum_{k=0}^{m} \binom{-m+k-1}{k} \nabla^k f_n. \qquad (2.35)$$

*Proof*

The inductive proof is similar to that of Lemma 2.4. Show that Eqn (2.35) is true for $m = 0$, assume it true for $m = 0$,

§2.8]  NEWTON'S BACKWARD DIFFERENCE FORMULA  47

..., $p - 1$, and hence show that it is true for $m = p$, using the results

$$f_{n-p+1} - f_{n-p} = \nabla f_{n-p+1},$$

and

$$\binom{-p+k}{k} - \binom{-p+k-1}{k-1} = \binom{-p+k-1}{k}$$

by Lemma 2.3. ∎

We then have the following theorem.

### THEOREM 2.9

With the same notation as in Theorem 2.8, except for $s$,

$$p_n(x) = \sum_{k=0}^{n} \binom{s+k-1}{k} \nabla^k f_n, \qquad (2.36)$$

where in Eqn (2.36)

$$s = \frac{(x - x_n)}{h}.$$

*Proof*

The proof is similar to that of Theorem 2.8.
Let $P_k(s)$ be defined by

$$P_k(s) = p_k(x) \quad (k = 0, \ldots, n)$$

where $p_k(x)$ interpolates $f$ on $\{x_n, x_{n-1}, \ldots, x_{n-k}\}$.

Then if Eqn (2.36) is true we have

$$P_0(s) = f_n \qquad (2.37a)$$

$$P_k(s) = P_{k-1}(s) + \binom{s+k-1}{k} \nabla^k f_n$$
$$(k = 1, \ldots, n). \quad (2.37b)$$

To establish Eqn (2.37b) by induction we first show it to be true for $k = 1$ by showing that $P_1(s)$ given by Eqn (2.37b) interpolates $f$ on $\{x_n, x_{n-1}\}$. We then assume Eqn (2.37b) true for $k = 1, \ldots, q - 1$, define $G_q(s)$ by

$$G_q(s) = P_{q-1}(s) + \binom{s+q-1}{q} \nabla^q f_n,$$

and show that
$$G_q(-r) = f_{n-r} \quad (r = 0, \ldots, q - 1).$$
Then
$$G_q(-q) = P_{q-1}(-q) + \binom{-1}{q}\nabla^q f_n = \sum_{k=0}^{q} \binom{-q+k-1}{k}\nabla^k f_n$$
$$= f_{n-p}, \quad \text{by Lemma 2.5.}$$

The remainder of the proof is similar to that of Theorem 2.8. ∎

From Theorem 2.9 we have the following algorithm for constructing Newton's backward difference form of the interpolating polynomial $p_n(x)$.

*Algorithm 2.2*

1. Set $s = (x - x_n)/h$ and $P_0(s) = f_n$.
2. For $k = 1, \ldots, n$ generate the $P_k(s)$ from
$$P_k(s) = P_{k-1}(s) + \binom{s+k-1}{k}\nabla^k f_n. \quad \blacksquare$$

*Example 2.11*

With $f$ tabulated in Table 2.6 let us estimate $f(0\cdot 5)$ using the tabular points $k = 0, \ldots, 3$.

We have $n = 3$, $x = 0\cdot 5$, $x_n = 0\cdot 6$, $h = 0\cdot 2$,
$$s = (x - x_n)/h = -\tfrac{1}{2}.$$
Then
$$P_0(s) = f_n = 0\cdot 3846,$$
$$P_1(s) = P_0(s) + \binom{s}{1}\nabla f_n = 0\cdot 3846 + \frac{0\cdot 0321}{2} = 0\cdot 4011,$$
$$P_2(s) = P_1(s) + \binom{s+1}{2}\nabla^2 f_n = 0\cdot 4011 - \frac{0\cdot 0057}{8} = 0\cdot 4004,$$
$$P_3(s) = P_2(s) + \binom{s+2}{3}\nabla^3 f_n = 0\cdot 4004 + \frac{0\cdot 0019}{16} = 0\cdot 4005.$$

The exact value of $f(0\cdot 5)$ is $0\cdot 4$.

To obtain a bound for the truncation error in Newton's

backward difference formula we again use Eqn (2.14) in the form

$$f(x) - p_n(x) = \prod_{k=0}^{n} (x - x_{n-k}) f^{(n+1)}(\xi)/(n+1)!$$

With
$$x = x_n + sh, \quad x_{n-k} = x_n - kh \quad (k = 0, \ldots, n).$$

Then we obtain

$$f(x) - P_n(s) = \binom{s+n}{n+1} h^{n+1} f^{(n+1)}(\xi), \qquad (2.38)$$

where
$$\min \{x_0 \ldots x_n, x\} < \xi < \max \{x_0 \ldots x_n, x\}.$$

*Example 2.12*

For Example 2.11 we have $s = -\tfrac{1}{2}$, $n = 3$, $h = 0.2$. So

$$|f(0.5) - p_3(0.5)| = \left|\binom{5/2}{4}(0.2)^4 f^{(4)}(\xi)\right|$$

$$\leqslant \binom{5/2}{4}(0.2)^4 \tfrac{3}{4} < 0.0012$$

whereas the actual error is 0.0005.

## Tutorial Examples

In the following problems, unless stated otherwise, $p_n$ denotes the interpolating polynomial for $f$ on the set of distinct points $\{x_k : k = 0, \ldots, n\}$.

1. Do there exist functions $f$ for which $p_n$ has degree less than $n$?

2. Using Lagrange interpolation construct $p_3$ for $f$ on $\{0, 1, 2, 3\}$ where
$$f(x) = x^2 + 2x + 1 \quad (0 \leqslant x \leqslant 3).$$

3. Apply the ideas contained in Section 2.5 to any tables which are available.

4. Obtain a bound for the total error due to truncation and rounding for linear interpolation in Example 2.4.

5. Let
$$x_k = x_0 + kh \quad (k = 1, 2, 3) \quad (h > 0),$$
and let
$$|f^{(4)}(x)| \leq M \quad (x_0 \leq x \leq x_3).$$

Show that for $x_1 \leq x \leq x_2$ the truncation error $e_T$ for $p_4$ has greatest magnitude at $x = (x_1 + x_2)/2$ and hence that
$$|e_T| \leq \frac{3Mh^4}{128} \quad (x_1 \leq x \leq x_2).$$

If each tabular value of $f$ has rounding error of magnitude at most s obtain a bound for the rounding error $e_R$ in $p_4$, on $[x_1, x_2]$.

6. Use Lagrange interpolation to construct $p_2$ for sinh $x$ on {1·90, 2·00, 2·10} using the following data.

| $x$ | 1·80 | 1·90 | 2·00 | 2·10 | 2·20 |
|---|---|---|---|---|---|
| sinh $x$ | 2·94217 | 3·26816 | 3·62686 | 4·02186 | 4·45711 |

Estimate the value of sinh 1·95, and obtain bounds for the truncation error and rounding error in $p_2$ (1·95).

7. Construct a forward difference table for the function $S$ of $n$ from the following data.

| $n$ | 0 | 1 | 2 | 3 | 4 | 5 |
|---|---|---|---|---|---|---|
| $S(n)$ | 0 | 1 | 9 | 36 | 100 | 225 |

From the difference table deduce the form of the function $S$. By using the constancy of the fourth-order differences compute the value of $S(6)$ without evaluating the formula for $S$ at $n = 6$. Check your results by noting that $S(n)$ is the sum of the cubes of the first $n$ non-negative integers.

8. Construct the forward difference table for the function $f$ from the following data at uniformly spaced points $x_k$,

$$f(x_k) = \begin{cases} 0 & (k = 0, \ldots, n-1) \\ \varepsilon & (k = n) \\ 0 & (k = n+1, \ldots, N), \end{cases}$$

taking $n = 3$, $N = 6$. By inspection, or by using Eqn (2.22) and noting that the only non-zero term in the sum is that for which $k = m + r - n$, verify that the non-zero terms of the difference table are given by

$$\Delta^m f_r = \binom{m}{m+r-n}(-1)^{m+r-n}\varepsilon.$$

9. The following is a table of values of a polynomial of degree five and contains a single error. Locate and correct this error using the result of Tutorial Example 2.8.

| $x$ | 0 | 1 | 2 | 3 | 4 | 5 | 6 |
|---|---|---|---|---|---|---|---|
| $f(x)$ | 1 | 2 | 33 | 254 | 1025 | 3126 | 7777 |

10. From the data in Tutorial Example 2.6 estimate the value of sinh 1.85 using Newton's forward difference formula, and obtain a bound for the truncation error.

11. From the data in Tutorial Example 2.6 estimate the value of sinh 2.15 using Newton's backward difference formula and obtain a bound on the truncation error. Obtain a bound for the rounding error in your estimate due to the rounding errors in the data. [*Hint:* Use Eqn (2.27).]

# CHAPTER 3

# Numerical Differentiation and Integration

## 3.1 Introduction

In this chapter we consider how to estimate the numerical value of a derivative and of a definite integral of a function $f$ from a table of values of $f$; that is, we consider numerical differentiation and integration.

Let a function $f$ be defined on $[a, b]$ and tabulated on a set $\{x_k: k = 0, \ldots, n\}$ of points in $[a, b]$ which need not be uniformly spaced, and let $f$ be differentiable and integrable on $[a, b]$. To estimate the numerical value of $f^{(m)}(x)$ where $x$ is a given point in $[a, b]$ and $m$ is a positive integer we could evaluate $p_n^{(m)}(x)$ where $p_n$ is the interpolating polynomial for $f$ on $\{x_k: k = 0, \ldots, n\}$. Even when $m$ is very much less than $n$ we should not expect this procedure to give a very accurate estimate of $f^{(m)}(x)$ in general. For example, it is intuitively obvious from Fig. 2.1 that even if $p_n(x)$ is a good approximation to $f(x)$ for $a \leqslant x \leqslant b$, $p_n^{(1)}(x)$ may be a bad approximation to $f^{(1)}(x)$. Also, in general, estimates of $f^{(m)}(x)$ obtained by differentiating $p_n(x)$ can be very sensitive to rounding errors in the tabular values of $f$ in addition to having large truncation errors. For these reasons, if it is impossible to avoid numerical differentiation altogether, care should be taken to obtain reliable bounds on both the truncation and rounding errors.

To estimate the numerical value of $\int_a^b f(x)\,dx$ we could evaluate $\int_a^b p_n(x)\,dx$. It is intuitively clear from Fig. 2.1 that this procedure is likely to have a small truncation error if $f$ is well approximated by $p_n$ on $[a, b]$, for in general the graph of $p_n$ oscillates about that of $f$ and the errors in the estimate of the integral due to the positive fluctuations of $p_n(x)$ relative to $f(x)$ would be expected to cancel the errors due to the negative fluctuations to some extent. If, however, a single interpolating polynomial $p_n$ is used to approximate $f$ on the whole interval $[a, b]$, then in general $n$ will have to be large to obtain reasonable accuracy. This leads to estimates of $\int_a^b f(x)\,dx$ with unacceptably large rounding errors. Rounding error can be reduced by dividing $[a, b]$ into a number of subintervals on each of which $f$ can be adequately approximated by an interpolating polynomial of low degree. Numerical integration using interpolating polynomials to approximate $f$ is sometimes called *interpolatory quadrature*.

## 3.2 Numerical Differentiation using the Interpolating Polynomial

From Theorem 2.2 we have, with the notation previously defined,

$$p_n(x) = \sum_{k=0}^{n} L_k(x) f_k,$$

whence

$$p_n^{(1)}(x) = \sum_{k=0}^{n} L_n^{(1)}(x) f_k. \qquad (3.1)$$

Taking $p_n^{(1)}(x)$ as an approximation to $f^{(1)}(x)$, Eqn (3.1) yields, for each value of $n$, a formula giving an estimate of $f^{(1)}(x)$. In practice, the most useful formulae are obtained when the $x_k$ are uniformly spaced. It is then convenient to obtain the formulae by using one of the forms of the interpolating polynomial specifically designed for uniformly spaced interpolating points. In particular we shall use Newton's forward difference formula.

Using the notation of Theorem 2.8 we have, since $p_n(x) = P_n(s)$ and $x = x_0 + sh$,

$$p_n^{(1)}(x) = \frac{d}{ds} P_n(s) \frac{ds}{dx} = \frac{1}{h} \frac{dP_n(s)}{ds}.$$

Hence
$$p_n^{(1)}(x) = \frac{1}{h} \sum_{k=0}^{n} \frac{d}{ds}\binom{s}{k}\Delta^k f_0 \quad (a \leqslant x \leqslant b). \tag{3.2}$$

The most useful results are obtained when $x = x_m$ ($0 \leqslant m \leqslant n$), in which case we have

$$p_n^{(1)}(x_m) = \frac{1}{h} \sum_{k=0}^{n} \frac{d}{ds}\binom{m}{k}\Delta^k f_0. \tag{3.3}$$

In Eqn (3.3),
$$\frac{d}{ds}\binom{m}{k} = \left[\frac{d}{ds}\binom{s}{k}\right]_{s=m}.$$

In Eqn (3.3) take $m = 0$ and $n = 1$. Then since

$$\frac{d}{ds}\binom{s}{0} = \frac{d}{ds}(1) = 0, \quad \frac{d}{ds}\binom{s}{1} = \frac{d}{ds}(s) = 1,$$

we have
$$p_1^{(1)}(x_0) = \frac{1}{h}\Delta f_0 = \frac{(f_1 - f_0)}{h}.$$

Hence we obtain, neglecting the truncation error, the differentiation formula

$$f^{(1)}(x_0) \approx \frac{(f_1 - f_0)}{h} \tag{3.4a}$$

The geometrical interpretation of Eqn (3.4a) is shown in Fig. 3.1, from which it is seen that the slope $f^{(1)}(x_0)$ of the tangent $P_0T$ to the graph of $f(x)$ at $x = x_0$ is approximated by the slope of the chord $P_0P_1$. Alternatively since the chord $P_0P_1$ is the graph of the interpolating polynomial of degree unity for $f$ on $\{x_0, x_1\}$ we see that Eqn (3.4a) is obtained by differentiating the linear interpolating polynomial.

In Eqn (3.3) take $m = 1$ and $n = 2$. Then since

$$\frac{d}{ds}\binom{s}{2} = \frac{d}{ds}\frac{s(s-1)}{2} = s - \frac{1}{2},$$

§3.2] DIFFERENTIATION BY INTERPOLATING POLYNOMIAL 55

we have, with $s = m = 1$,

$$p_2^{(1)}(x_1) = \frac{1}{h}[\Delta f_0 + \tfrac{1}{2}\Delta^2 f_0] = \frac{(f_2 - f_0)}{2h}.$$

Hence we obtain, neglecting the truncation error, the differentiation formula

$$f^{(1)}(x_1) \approx \frac{(f_2 - f_0)}{2h}. \tag{3.5a}$$

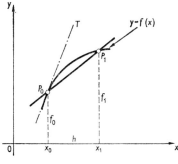

FIG 3.1

The geometrical interpretation of Eqn (3.5a) is shown in Fig. 3.2, from which it is seen that the slope of $f^{(1)}(x_1)$ of the tangent $P_1 T$ to the graph of $f(x)$ at $x = x_1$ is approximated by the slope of the chord $P_0 P_2$.

Since the labelling of the ordinates $f_k$ is immaterial provided that consecutive ordinates are labelled consecutively, we have

$$f^{(1)}(x_m) \approx \frac{(f_{m+1} - f_m)}{h} \quad (m = 0, 1, 2, \ldots), \tag{3.4b}$$

and

$$f^{(1)}(x_m) \approx \frac{(f_{m+1} - f_{m-1})}{2h} \quad (m = 1, 2, \ldots). \tag{3.5b}$$

By making other choices of $m$ and $n$ in Eqn (3.3) many other differentiation formulae can be obtained. For example, by taking $m = 0$ and $n = 2$ we obtain

$$f^{(1)}(x_0) \approx \frac{1}{2h}(-3f_0 + 4f_1 - f_2), \tag{3.6a}$$

and more generally

$$f^{(1)}(x_m) \approx \frac{1}{2h}(-3f_m + 4f_{m+1} - f_{m+2}) \quad (m = 0, 1, \ldots). \tag{3.6b}$$

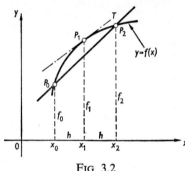

Fig 3.2

*Example 3.1*

We shall estimate $f^{(1)}(0\cdot 2)$ from the data in Table 3.1 using formulae (3.4), (3.5), and (3.6).

| $x$ | 0·1 | 0·2 | 0·3 | 0·4 |
|---|---|---|---|---|
| $f(x)$ | 0·0001 | 0·0016 | 0·0081 | 0·0256 |

Table 3.1

Using formula (3.4) take $x_0 = 0\cdot 2$, $x_1 = 0\cdot 3$, $h = 0\cdot 1$, whence $f^{(1)}(0\cdot 2) \approx 0\cdot 065$. Using formula (3.5) take $x_0 = 0\cdot 1$, $x_1 = 0\cdot 2$, $x_2 = 0\cdot 3$, $h = 0\cdot 1$, whence $f^{(1)}(0\cdot 2) \approx 0\cdot 04$. Using formula (3.6) take $x_0 = 0\cdot 2$, $x_1 = 0\cdot 3$, $x_2 = 0\cdot 4$, $h = 0\cdot 1$, whence $f^{(1)}(0\cdot 2) \approx 0\cdot 01$. The exact value of $f^{(1)}(0\cdot 2)$ is $0\cdot 032$ since $f(x) = x^4$. The best result is obtained using formula (3.5) as would be expected intuitively from an examination of the graph of $x^4$.

### 3.3 A Truncation Error Formula for Numerical Differentiation

Using Lemma 2.1 it is easy to obtain a formula for the truncation error $f^{(1)}(x_m) - p_n^{(1)}(x_m)$ resulting from the use of Eqn (3.1) to estimate $f^{(1)}(x_m)(0 \leqslant m \leqslant n)$. We have the following theorem.

### THEOREM 3.1

Let $f^{(n+1)}(x)$ be continuous on $[a, b]$ and let $\{x_k: k = 0, \ldots, n\}$ be any $(n + 1)$ distinct points on $[a, b]$. Let $p_n(x)$ be the interpolating polynomial for $f$ on $\{x_k: k = 0, \ldots, n\}$. Then

$$f^{(1)}(x_m) - p_n^{(1)}(x_m) = \prod_{\substack{k=0 \\ k \neq m}}^{n} (x_m - x_k) \frac{f^{(n+1)}(\xi_m)}{(n+1)!}$$

$$(m = 0, \ldots, n) \quad (3.7)$$

where

$$\min \{x_0, \ldots, x_n\} < \xi_m < \max \{x_0, \ldots, x_n\}.$$

*Proof*

Let
$$F(x) = \prod_{k=0}^{n} (x - x_k).$$

Then by Lemma 2.1,

$$F^{(1)}(x_m) = \prod_{\substack{k=0 \\ k \neq m}}^{n} (x_m - x_k),$$

so $F^{(1)}(x_m) \neq 0$. Let

$$\mu = \frac{f^{(1)}(x_m) - p_n^{(1)}(x_m)}{F^{(1)}(x_m)},$$

and let

$$G(x) = f(x) - p_n(x) - \mu F(x).$$

Then by definition of $p_n(x)$ and $F(x)$,

$$G(x_k) = 0 \quad (k = 0, \ldots, n).$$

Hence by Lemma 2.2, there is a number $\xi_m$ such that

$$G^{(n+1)}(\xi_m) = 0$$

and $\quad \min \{x_0, \ldots, x_n\} < \xi_m < \max \{x, \ldots, x_n\}.$

Hence since
$$G^{(n+1)}(x) = f^{(n+1)}(x) - \mu(n+1)!,$$
we have
$$\mu = \frac{f^{(n+1)}(\xi_m)}{(n+1)!},$$
whence Eqn (3.7) is established. ∎

We note that in general Eqn (3.7) cannot be used to obtain the *exact* value of the truncation error, but it can be used to obtain a bound for the magnitude of the truncation error as illustrated in Example 3.2. Firstly we apply Theorem 3.1 to formulae (3.4), (3.5), and (3.6).

For formula (3.4a), $m = 0$, $n = 1$, and we have
$$f^{(1)}(x_0) - p_1^{(1)}(x_0) = \frac{(x_0 - x_1)}{2} f^{(2)}(\xi_0)$$
$$= -\frac{h}{2} f^{(2)}(\xi_0) \quad (x_0 < \xi_0 < x_1). \quad (3.8)$$

For formula (3.5a), $m = 1$, $n = 2$, and we have
$$f^{(1)}(x_1) - p_2^{(1)}(x_1) = \frac{(x_1 - x_0)(x_1 - x_2)}{6} f^{(3)}(\xi_1)$$
$$= \frac{h^2}{6} f^{(3)}(\xi_1) \quad (x_0 < \xi_1 < x_2). \quad (3.9)$$

For formula (3.6a), $m = 0$, $n = 2$ and we have
$$f^{(1)}(x_0) - p_2^{(1)}(x_0) = \frac{(x_0 - x_1)(x_0 - x_2)}{6} f^{(3)}(\xi_0)$$
$$= \frac{h^2}{3} f^{(3)}(\xi_0) \quad (x_0 < \xi_0 < x_2). \quad (3.10)$$

*Example 3.2*

Let us apply Eqns (3.8), (3.9), and (3.10) to Example 3.1. We have $f(x) = x^4$, $f^{(1)}(x) = 4x^3$, $f^{(2)}(x) = 12x^2$, $f^{(3)}(x) = 24x$. Hence
$$|f^{(2)}(x)| \leqslant 1 \cdot 08 \quad (0 \cdot 2 \leqslant x \leqslant 0 \cdot 3),$$
$$|f^{(3)}(x)| \leqslant 7 \cdot 2 \quad (0 \cdot 1 \leqslant x \leqslant 0 \cdot 3),$$
$$|f^{(3)}(x)| \leqslant 9 \cdot 6 \quad (0 \cdot 2 \leqslant x \leqslant 0 \cdot 4).$$

§3.3] TRUNCATION ERROR FORMULA FOR DIFFERENTIATION 59

So the magnitudes of the truncation errors $E_{T1}$, $E_{T2}$, and $E_{T3}$ obtained by using formulae (3.4), (3.5), and (3.6) respectively are bounded according to

$$|E_{T1}| = \left|\frac{h}{2}f^{(2)}(\xi_0)\right| \leqslant \frac{0 \cdot 1}{2} \times 1 \cdot 08 = 0 \cdot 054,$$

$$|E_{T1}| = \left|\frac{h^2}{6}f^{(3)}(\xi_1)\right| \leqslant \frac{0 \cdot 01}{6} \times 7 \cdot 2 = 0 \cdot 012,$$

$$|E_{T3}| = \left|\frac{h^2}{3}f^{(3)}(\xi_0)\right| \leqslant \frac{0 \cdot 01}{3} \times 9 \cdot 6 = 0 \cdot 032.$$

The magnitudes of the actual errors obtained were 0·033, 0·008, and 0·022 respectively. Note the superior accuracy of formula (3.5).

If $q$ is a function of $h$ such that for $h$ sufficiently small

$$|q(h)| \leqslant Ah^k$$

where $A$ is a constant independent of $h$, then we say that *q is of order $h^k$* and we write

$$q = O(h^k).$$

The truncation errors in formulae (3.4), (3.5), and (3.6) are therefore $O(h)$, $O(h^2)$, and $O(h^2)$ respectively, and we refer to formulae (3.4) and (3.5) as first- and second-order formulae respectively.

## 3.4 The Taylor Expansion Method for Numerical Differentiation

We now consider a method by means of which any differentiation formula of specified type can be obtained, together with a bound on the truncation error. The general idea will be illustrated by deriving formula (3.5).

It is required to obtain a formula for $f^{(1)}(x_1)$ of the form

$$f^{(1)}(x_1) = \frac{1}{h}(\alpha_1 f_0 + \alpha_2 f_1 + \alpha_3 f_2) + E_T \qquad (3.11)$$

where $\alpha_1$, $\alpha_2$, and $\alpha_3$ are coefficients to be determined. Write $f_k^{(m)} = f^{(m)}(x_k)$. Then if $f^{(3)}(x)$ exists in $(x_0, x_2)$, we have

$$f_0 = f(x_1 - h) = f_1 - hf_1^{(1)} + \frac{h^2}{2}f_1^{(2)} - \frac{h^3}{6}f^{(3)}(\xi_1)$$
$$(x_0 < \xi_1 < x_1), \quad (3.12)$$
$$f_2 = f(x_1 + h) = f_1 + hf_1^{(1)} + \frac{h^2}{2}f_1^{(2)} + \frac{h^3}{6}f^{(3)}(\xi_2)$$
$$(x_1 < \xi_2 < x_2). \quad (3.13)$$

Substituting into (3.11) from (3.12) and (3.13) we obtain

$$f_1^{(1)} = \frac{1}{h}(\alpha_1 + \alpha_2 + \alpha_3)f_1$$
$$+ (-\alpha_1 + \alpha_3)f_1^{(1)} + \frac{h}{2}(\alpha_1 + \alpha_3)f_1^{(2)}$$
$$+ \frac{h^2}{6}[\alpha_3 f^{(3)}(\xi_2) - \alpha_1 f^{(3)}(\xi_1)] + E_T. \quad (3.14)$$

If Eqn (3.14) is to be identically true for all functions $f$ for which $f^{(3)}$ exists in $(x_0, x_2)$ we can equate coefficients of $f_1, f_1^{(1)},$ and $f_1^{(2)}$ to obtain

$$\alpha_1 + \alpha_2 + \alpha_3 = 0,$$
$$-\alpha_1 \phantom{{}+\alpha_2} + \alpha_3 = 1,$$
$$\alpha_1 \phantom{{}+\alpha_2} + \alpha_3 = 0,$$

whence $\alpha_1 = -\frac{1}{2}$, $\alpha_2 = 0$, $\alpha_3 = \frac{1}{2}$.

Substituting into Eqn (3.14) we obtain

$$E_T = -\frac{h^2}{12}[f^{(3)}(\xi_1) + f^{(3)}(\xi_2)].$$

If $f^{(3)}$ is continuous on $[x_0, x_2]$ we deduce the existence of a number $\xi$ in $(\xi_1, \xi_2)$ such that

$$E_T = -\frac{h^2}{6}f^{(3)}(\xi).$$

Hence we recover the result

$$f^{(1)}(x_1) = \frac{(f_2 - f_0)}{2h} - \frac{h^2}{6}f^{(3)}(\xi) \quad (x_0 < \xi < x_2). \quad (3.15)$$

Note that if $f$ is a polynomial of degree at most 2, the trun-

## §3.4] DIFFERENTIATION BY TAYLOR EXPANSION METHOD

cation error is zero and the formula corresponding to Eqn (3.15) is exact. If a differentiation formula is exact for a polynomial of degree at most $m$, then the formula is said to have a *degree of precision m*.

Formulae for higher derivatives of $f$ can be obtained by repeatedly differentiating the interpolating polynomial for $f$ on a given set of interpolating points, or by the Taylor expansion method as in the derivation of Eqn (3.15). Since it is not possible to obtain a simple error formula of the same type as Eqn (3.7) for derivatives of order greater than unity by differentiating the interpolating polynomial we consider instead the Taylor expansion method. We shall obtain a formula for $f^{(2)}(x_1)$ of the form

$$f^{(2)}(x_1) = \frac{1}{h^2}(\alpha_1 f_0 + \alpha_2 f_1 + \alpha_3 f_2) + E_T. \quad (3.16)$$

Provided $f^{(4)}(x)$ exists in $(x_0, x_2)$ we have

$$f_0 = f(x_1 - h) = f_1 - hf_1^{(1)} + \frac{h^2}{2}f_1^{(2)} - \frac{h^3}{6}f_1^{(3)} + \frac{h^4}{24}f^{(4)}(\xi_1)$$
$$(x_0 < \xi_1 < x_1),$$

$$f_2 = f(x_1 + h) = f_1 + hf_1^{(1)} + \frac{h^2}{2}f_1^{(2)} + \frac{h^3}{6}f_1^{(3)} + \frac{h^4}{24}f^{(4)}(\xi_2)$$
$$(x_1 < \xi_2 < x_2).$$

Substituting these results into Eqn (3.16) we obtain

$$f_1^{(2)} = \frac{1}{h^2}(\alpha_1 + \alpha_2 + \alpha_3)f_1 + \frac{1}{h}(-\alpha_1 + \alpha_3)f_1^{(1)}$$
$$+ \frac{1}{2}(\alpha_1 + \alpha_3)f_1^{(2)} + \frac{h}{6}(-\alpha_1 + \alpha_3)f_1^{(3)}$$
$$+ \frac{h^2}{24}[\alpha_1 f^{(4)}(\xi_1) + \alpha_3 f^{(4)}(\xi_2)] + E_T. \quad (3.17)$$

Equating coefficients of $f_1, f_1^{(1)}$, and $f_1^{(2)}$ we obtain

$$\begin{aligned} \alpha_1 + \alpha_2 + \alpha_3 &= 0 \\ -\alpha_1 \quad\quad\; + \alpha_3 &= 0 \\ \alpha_1 \quad\quad\; + \alpha_3 &= 2, \end{aligned}$$

whence $\alpha_1 = 1$, $\alpha_2 = -2$, and $\alpha_3 = 1$.

Substituting these values into Eqn (3.17) we obtain

$$E_T = \frac{h^2}{24}[f^{(4)}(\xi_1) + f^{(4)}(\xi_2)].$$

If $f^{(4)}$ is continuous on $[x_0, x_2]$ then

$$E_T = \frac{h^2}{12} f^{(4)}(\xi) \quad (x_0 < \xi < x_2).$$

Hence we obtain the formula

$$f^{(2)}(x_1) = \frac{1}{h^2}(f_0 - 2f_1 + f_2) - \frac{h^2}{12} f^{(4)}(\xi)$$
$$(x_0 < \xi < x_2). \quad (3.18)$$

*Example 3.3*

(1) Using Eqn (3.18) to estimate $f^{(2)}(0.2)$ in Example 3.1 we have $n = 2$, $x_0 = 0.1$, $x_1 = 0.2$, $x_2 = 0.3$, $h = 0.1$, whence

$$f^{(2)}(0.2) \approx \frac{1}{0.01}(0.0001 - 2 \times 0.0016 + 0.0081) = 0.5.$$

To obtain a bound for the truncation error we have, since $f^{(4)}(x) = 24$

$$|E_T| \leqslant \frac{0.01}{12} \times 24 = 0.02.$$

The exact value of $f^{(2)}(0.2)$ is 0.48 so the error actually obtained is also 0.02 in this case.

## 3.5 Rounding Error in Numerical Differentiation

Suppose that it is required to estimate $f^{(1)}(x_m)$ from a formula of the form

$$f^{(1)}(x_m) = \frac{1}{h} \sum_{k=0}^{n} \alpha_k f_k + E_T.$$

Suppose also that approximate values $\hat{f}_k$ ($k = 0, \ldots, n$) of the exact tabular values $f_k$ ($k = 0, \ldots, n$) are known. Then the estimate $\hat{f}_m^{(1)}$ of $f^{(1)}(x_m)$ actually computed is given by

§3.5] ROUNDING ERROR IN NUMERICAL DIFFERENTIATION

$$\hat{f}_m^{(1)} = \frac{1}{h} \sum_{k=0}^{n} \alpha_k \hat{f}_k.$$

If

$$f_m^{(1)} = \frac{1}{h} \sum_{k=0}^{n} \alpha_k f_k, \tag{3.19}$$

then the rounding error $E_R$ is given by

$$E_R = f_m^{(1)} - \hat{f}_m^{(1)} = \frac{1}{h} \sum_{k=0}^{n} \alpha_k (f_k - \hat{f}_k).$$

So

$$|E_R| \leqslant \frac{1}{h} \sum_{k=0}^{n} |\alpha_k| \cdot |f_k - \hat{f}_k|. \tag{3.20}$$

If the difference between $f_k$ and $\hat{f}_k$ ($k = 0, \ldots, n$) is due to rounding error then we have

$$|f_k - \hat{f}_k| \leqslant \varepsilon \quad (k = 0, \ldots, n)$$

where the value of $\varepsilon$ depends upon the number of decimal places which are correct. For example, if $f_k$ is known correct to $q$D then $\varepsilon \leqslant 10^{-q}/2$.

Hence from Eqn (3.20)

$$|E_R| \leqslant \frac{\varepsilon}{h} \cdot \sum_{k=0}^{n} |\alpha_k|. \tag{3.21}$$

The total error $E$ in $\hat{f}_m^{(1)}$ regarded as an estimate of $f^{(1)}(x_m)$ is then bounded by

$$|E| \leqslant |E_R| + |E_T|.$$

Clearly the preceding argument is valid when considering derivatives of any order.

For formula (3.4), $\alpha_0 = -1$, $\alpha_1 = 1$, $n = 1$ and

$$|E_R| \leqslant \frac{2\varepsilon}{h}.$$

For formula (3.5), $\alpha_0 = -\frac{1}{2}$, $\alpha_1 = 0$, $\alpha_2 = \frac{1}{2}$, $n = 2$, and

$$|E_R| \leqslant \frac{\varepsilon}{h}.$$

For formula (3.6), $\alpha_0 = -\frac{3}{2}$, $\alpha_1 = 2$, $\alpha_2 = -\frac{1}{2}$, $n = 2$, and

$$|E_R| \leqslant \frac{4\varepsilon}{h}.$$

Comparing formulae (3.4), (3.5), and (3.6) by examining their truncation and rounding errors it is clear that formula (3.5) has some very desirable characteristics.

By examining Eqn (3.21) we note that the estimate $f_m^{(1)}$ of $f^{(1)}(x_m)$ is very sensitive to rounding errors in the $f_k$ owing to the factor $1/h$. Whereas decreasing $h$ causes $|E_T|$ to decrease, $|E_R|$ actually *increases* as $h$ is decreased. This means that the accuracy attainable using formulae of the type (3.19) is limited in the presence of rounding error and that there is an optimum value of $h$ which yields the smallest value of $E$. This is illustrated by the following example.

*Example 3.4*

Suppose that it is required to estimate $f^{(2)}(x)$ using Eqn (3.18). We shall obtain the interval $h$ at which $f$ must be tabulated to obtain the best results. If $f^{(4)}(x)$ is continuous on an interval $[a, b]$ and

$$|f^{(4)}(x)| \leqslant M \quad (a \leqslant x \leqslant b)$$

then

$$|E_T| \leqslant \frac{Mh^2}{12}.$$

If the magnitude of the rounding error in the tabular values of $f$ is at most $\varepsilon$, then

$$|E_R| \leqslant \frac{4\varepsilon}{h^2}.$$

Hence the total error $E$ in the formula corresponding to Eqn (3.18) is bounded by

$$|E| \leqslant \frac{4\varepsilon}{h^2} + \frac{Mh^2}{12} = y(h).$$

The value of $h$ which corresponds to the minimum value of $y$ satisfies $y^{(1)}(h) = 0$, whence

$$h = \left(\frac{48\varepsilon}{M}\right)^{1/4}.$$

Similar considerations can be applied to the other differentiation formulae.

## 3.6 Numerical Integration using the Interpolating Polynomial

From Theorem 2.2 with the notation previously defined we have

$$p_n(x) = \sum_{k=0}^{n} L_k(x) f_k,$$

whence

$$\int_a^b p_n(x) \, dx = \sum_{k=0}^{n} f_k \cdot \int_a^b L_k(x) \, dx. \tag{3.22}$$

The coefficients $\{\alpha_k : k = 0, \ldots, n\}$ defined by

$$\alpha_k = \int_a^b L_k(x) \, dx \quad (k = 0, \ldots, n)$$

depend only upon the interpolating points $\{x_k : k = 0, \ldots, n\}$ and the interval $[a, b]$, but do not depend upon $f$. If

$$I(a, b) = \int_a^b f(x) \, dx, \tag{3.23}$$

then an estimate of the value of $I$ is provided by $\int_a^b p_n(x) \, dx$, and so

$$I(a, b) = \sum_{k=0}^{n} \alpha_k f_k + E_T, \tag{3.24}$$

where $E_T$ is the truncation error associated with the *quadrature formula*

$$I(a, b) \approx \sum_{k=0}^{n} \alpha_k f_k.$$

The most useful quadrature formulae for subsequent

applications in this book are obtained when $n$ is small and the interpolating points are uniformly spaced.

It is then convenient to use, for example, Newton's forward difference form of the interpolating polynomial to obtain quadrature formulae. Using the notation of Theorem 2.8 we have, since $x = x_0 + sh$ and $p_n(x) = P_n(s)$,

$$\int_{x_0}^{x_m} p_n(x)\,dx = \int_0^m P_n(s)h\,ds = h\sum_{k=0}^n \Delta^k f_0 \int_0^m \binom{s}{k} ds. \quad (3.25)$$

We now consider some important special cases of this result.

(1) ($m = 1, n = 0$)

From (3.25) we obtain

$$\int_{x_0}^{x_1} p_0(x)\,dx = hf_0.$$

Using this result to approximate $I(x_0, x_1)$ we have the quadrature formula

$$I(x_0, x_1) \approx hf_0. \quad (3.26)$$

(2) ($m = 1, n = 1$)

From (3.25) we obtain

$$\int_{x_0}^{x_1} p_1(x)\,dx = \frac{h}{2}[f_0 + f_1].$$

Using this result to approximate $I(x_0, x_1)$ we have the quadrature formula

$$I(x_0, x_1) \approx \frac{h}{2}[f_0 + f_1]. \quad (3.27)$$

(3) ($m = 2, n = 2$)

From (3.25) we obtain

$$\int_{x_0}^{x_2} p_2(x)\,dx = \frac{h}{3}[f_0 + 4f_1 + f_2].$$

Using this result to approximate $I(x_0, x_2)$ we have the quadrature formula

$$I(x_0, x_2) \approx \frac{h}{3}[f_0 + 4f_1 + f_2]. \quad (3.28)$$

§3.6] INTEGRATION WITH INTERPOLATING POLYNOMIAL 67

The formulae (3.26), (3.27), and (3.28) have simple geometrical interpretations. If we approximate the area ABCD in Fig. 3.3 under the graph of $f(x)$ by the area of the rectangle *ABCN* we obtain formula (3.26) which for this reason is called the *rectangle rule*.

If we approximate the area *ABCD* under the graph of $f(x)$ by the area of the trapezium *ABCD* we obtain formula (3.27) which for this reason is called the *trapezium rule*. Formula

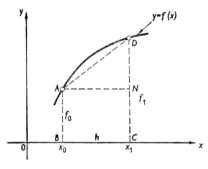

FIG. 3.3

(3.28) is called *Simpson's rule* and is obtained geometrically by approximating the area under the graph of $f(x)$ between $x = x_0$ and $x = x_2$ by the area under the parabola which passes through the points $(x_0, f_0)$, $(x_1, f_1)$, and $(x_2, f_2)$.

To apply formulae (3.26) and (3.27) to any finite interval of integration $[a, b]$ we choose any positive integer $n \geqslant 1$ and divide $[a, b]$ into $n$ segments each of length $h$, using points which we label $x_k$ so that

$$h = \frac{(b-a)}{n}, \quad x_0 = a, \quad x_k = x_0 + kh \quad (k = 0, \ldots, n).$$

Then we have

$$I(a, b) = \sum_{k=0}^{n-1} I(x_k, x_{k+1}).$$

Using the rectangle rule to approximate $I(x_k, x_{k+1})$ ($k = 0, \ldots, n-1$) we then obtain the quadrature formula

$$I(a, b) \approx h[f_0 + f_1 + \ldots + f_{n-1}]. \tag{3.29}$$

Using the trapezium rule to approximate $I(x_k, x_{k+1})$ ($k = 0, \ldots, n-1$) we obtain the quadrature formula

$$I(a, b) \approx \frac{h}{2}[f_0 + 2f_1 + 2f_2 + \ldots + 2f_{n-1} + f_n]. \tag{3.30}$$

To apply formula (3.28) we divide $[a, b]$ into an *even* number $2n$ of segments each of length $h$, so that

$$h = \frac{(b-a)}{2n}, \quad x_0 = a, \quad x_k = x_0 + kh \quad (k = 0, \ldots, 2n).$$

Then we have

$$I(a, b) = \sum_{k=0}^{2n-2} I(x_k, x_{k+2}),$$

whence

$$I(a, b) \approx \frac{h}{3}[f_0 + 4f_1 + 2f_2 + 4f_3 + \ldots + f_{2n}]. \tag{3.31}$$

*Example 3.5*

We shall estimate the value of $\int_1^2 f(x)\,dx$ from the data in Table 3.2 using formulae (3.29), (3.30), and (3.31) with various values of $h$.

(1) *Rectangle rule*

With $h = 0.2$,
$$\begin{aligned}I &\approx 0.2[1.0000000 + 0.8333333 + 0.7142857 \\ &\quad + 0.6250000 + 0.5555556] \\ &= 0.7456349.\end{aligned}$$

With $h = 0.1$,
$$\begin{aligned}I &\approx 0.1[1.0000000 + 0.9090909 + 0.8333333 + 0.7692308 \\ &\quad + 0.7142857 + 0.6666667 + 0.6250000 \\ &\quad + 0.5882353 + 0.5555556 + 0.5263158] \\ &= 0.7187714.\end{aligned}$$

§3.6]  INTEGRATION WITH INTERPOLATING POLYNOMIAL

| $x$ | $f(x)$ | $x$ | $f(x)$ |
|---|---|---|---|
| 1·00 | 1·0000000 | 1·55 | 0·6451613 |
| 1·05 | 0·9523809 | 1·60 | 0·6250000 |
| 1·10 | 0·9090909 | 1·65 | 0·6060606 |
| 1·15 | 0·8695652 | 1·70 | 0·5882353 |
| 1·20 | 0·8333333 | 1·75 | 0·5714286 |
| 1·25 | 0·8000000 | 1·80 | 0·5555556 |
| 1·30 | 0·7692308 | 1·85 | 0·5405405 |
| 1·35 | 0·7407407 | 1·90 | 0·5263158 |
| 1·40 | 0·7142857 | 1·95 | 0·5128205 |
| 1·45 | 0·6896552 | 2·00 | 0·5000000 |
| 1·50 | 0·6666667 | | |

TABLE 3.2

With $h = 0·05$, we obtain similarly, the estimate 0·7058034 for $I$.

(2) *Trapezium rule*

With $h = 0·2$,

$I \approx (0·2/2)[1·0000000 + 2(0·8333333 + 0·7142857$
$\quad + 0·6250000 + 0·5555556) + 0·5000000]$
$= 0·6956349.$

With $h = 0.1$,

$I \approx (0.1/2)[1.0000000 + 2(0.9090909 + 0.8333333$
$+ 0.7692308 + 0.7142857 + 0.6666667$
$+ 0.6250000 + 0.5882353 + 0.5555556$
$+ 0.5263158) + 0.5000000]$
$= 0.6937714.$

Similarly with $h = 0.05$, we obtain the estimate $0.6933034$ for $I$.

(3) *Simpson's rule*

With $h = 0.25$,
$I \approx (0.25/3)[1.0000000 + 4 \times 0.8000000 + 2 \times 0.6666667$
$+ 4 \times 0.5714286 + 0.5000000]$
$= 0.6932540.$

With $h = 0.1$,

$I \approx (0.1/3)[1.0000000 + 4 \times 0.9090909 + 2 \times 0.8333333$
$+ 4 \times 0.7692308 + 2 \times 0.7142857 + 4 \times 0.6666667$
$+ 2 \times 0.6250000 + 4 \times 0.5882353 + 2 \times 0.5555556$
$+ 4 \times 0.5263158 + 0.5000000]$
$= 0.6931502.$

With $h = 0.05$ we obtain, similarly, the estimate $0.6931473$ for $I$.

In Table 3.2 $f(x) = 1/x$ and so

$$I = \int_1^2 \frac{1}{x} \, dx = \log 2 = 0.6931472$$

correct to 7D.

From the preceding calculations we note that, at least for the example cited,

(a) for a given value of $h$, Simpson's rule is much more accurate than the trapezium rule, which is in turn more accurate than the rectangle rule;

(b) for a given accuracy, Simpson's rule requires less computational labour than the trapezium rule which in turn requires less computational labour than the rectangle rule.

## 3.7 Truncation Error in Interpolatory Quadrature

Let $f$ be defined on an interval $[c, d]$ and let $c \leqslant a < b \leqslant d$. Let $\{x_k : k = 0, \ldots, n\}$ be any $(n + 1)$ points in $[c, d]$. Then from Theorem 2.3, with $\phi_n(x)$ defined appropriately,

$$f(x) - p_n(x) = \frac{\phi_n(x) f^{(n+1)}(\xi(x))}{(n+1)!} \quad (c \leqslant x \leqslant d),$$

where

$$\min \{x, x_0, \ldots, x_n\} < \xi(x) < \max \{x, x_0, \ldots, x_n\}.$$

The truncation error $E_T$ defined by Eqn (3.24) is thus given by

$$E_T = \int_a^b [f(x) - p_n(x)] \, dx = \frac{1}{(n+1)!} \int_a^b \phi_n(x) f^{(n+1)}(\xi(x)) \, dx. \tag{3.32}$$

Now if $f^{(n+1)}(\xi(x))$ is continuous on $[a, b]$ then

$$\phi_n(x) f^{(n+1)}(\xi(x))$$

is integrable over $[a, b]$ and $f^{(n+1)}(\xi(x))$ is bounded on $[a, b]$. Noting that some of the interpolating points may not be in $[a, b]$, let

$$|f^{(n+1)}(x)| \leqslant M$$
$$(\min \{x_0, \ldots, x_n, a, b\} \leqslant x \leqslant \max \{x_0, \ldots, x_n, a, b\}).$$

Then

$$|E_T| \leqslant \frac{M}{(n+1)!} \int_a^b |\phi_n(x)| \, dx. \tag{3.33}$$

Equation (3.33) provides an upper bound for the truncation error (although not very sharp in some cases) in all interpolatory quadrature formulae. In general, $[a, b]$ contains several interpolating points so that $\phi_n(x)$ changes sign at least once in $(a, b)$. If the interpolating points have uniform spacing $h$ then with

$$x = x_0 + sh, \quad x_k = x_0 + kh$$

we have

$$\phi_n(x) = h^{n+1} \prod_{k=0}^{n} (s-k),$$

and we obtain

$$|E_T| \leq \frac{h^{n+2}}{(n+1)!} \cdot M \cdot \int_{s_a}^{s_b} \left| \prod_{k=0}^{n} (s-k) \right| ds \qquad (3.34)$$

where

$$s_a = \frac{(a-x_0)}{h}, \quad s_b = \frac{(b-x_0)}{h}.$$

We shall apply inequality (3.34) to the special cases (1), (2), and (3) of Eqn (3.25).

*Case (1)*

With $a = x_0$, $b = x_1$, $n = 0$ we have for the rectangle rule on $[x_0, x_1]$

$$|E_T| \leq h^2 M \int_0^1 s \, ds = \frac{Mh^2}{2}, \qquad (3.35)$$

where

$$M = \max_{x_0 \leq x \leq x_1} |f^{(1)}(x)|.$$

Noting that $\phi_0(x)$ does not change sign on $[x_0, x_1]$ we obtain for the rectangle rule on $[x_0\ x_1]$,

$$E_T = h^2 f^{(1)}(\xi_0) \int_0^1 s \, ds = \frac{h^2}{2} f^{(1)}(\xi_0) \quad (x_0 < \xi_0 < x_1),$$

provided that $f^{(1)}(x)$ is continuous on $[x_0, x_1]$.

Hence we have

$$\int_{x_0}^{x_1} f(x) \, dx = hf_0 + \frac{h^2}{2} f^{(1)}(\xi_0) \quad (x_0 < \xi_0 < x_1). \qquad (3.36)$$

For integration over $[a, b]$ we then have, with

$$h = (b-a)/n, \quad x_0 = a, \quad x_k = x_0 + kh \\ (k = 0, \ldots, n) \qquad (3.37)$$

§3.7] TRUNCATION ERROR IN INTERPOLATORY QUADRATURE 73

$$\int_a^b f(x)\,\mathrm{d}x = h\sum_{k=0}^{n-1} f_k + \frac{h^2}{2}\cdot\sum_{k=0}^{n-1} f^{(1)}(\xi_k) \quad (x_k < \xi_k < x_{k+1}).$$

Provided that $f^{(1)}(x)$ is continuous on $[a, b]$ we then have by Lemma 2.2

$$\int_a^b f(x)\,\mathrm{d}x = h\sum_{k=0}^{n-1} f_k + \frac{nh^2}{2}\cdot f^{(1)}(\xi) \quad (a < \xi < b). \qquad (3.38)$$

*Example 3.6*

Consider Example 3.5 (1). With $f(x) = 1/x$, $a = 1$, $b = 2$ we have

$$|f^{(1)}(x)| = \left|\frac{1}{x}\right| \leqslant 1 \quad (1 \leqslant x \leqslant 2).$$

With $h = 0.2$, the truncation error $E_T$ obtained using the rectangle rule to evaluate $\int_1^2 f(x)\,\mathrm{d}x$ is then bounded by

$$|E_T| \leqslant \frac{5 \times (0.2)^2 \times 1}{2} = 0.1.$$

This means that, *neglecting rounding error*, the use of the rectangle rule with $h = 0.2$ with the data in Table 3.2 yields an estimate of $\int_1^2 (1/x)\,\mathrm{d}x$ which is in error by at most $\pm 0.1$. Similarly, with $h = 0.1$,

$$|E_T| \leqslant \frac{10 \times (0.1)^2 \times 1}{2} = 0.05,$$

and with $h = 0.05$,

$$|E_T| \leqslant \frac{20 \times (0.05)^2 \times 1}{2} = 0.025.$$

*Case* (2)

With $a = x_0$, $b = x_1$, and $n = 1$ in the inequality (3.34) we obtain for the trapezium rule on $[x_0, x_1]$

$$|E_T| \leqslant \frac{Mh^3}{2}\cdot\int_0^1 |s(s-1)|\,\mathrm{d}s = \frac{Mh^3}{12}, \qquad (3.39)$$

where $M = \max\limits_{x_0 \leqslant x \leqslant x_1} |f^{(2)}(x)|$.

Noting that $\phi_1(s)$ does not change sign in $[x_0, x_1]$ we obtain for the trapezium rule on $[x_0, x_1]$

$$E_T = \frac{h^3}{2}f^{(2)}(\xi_0)\int_0^1 s(s-1)\,ds = -\frac{h^3}{12}f^{(2)}(\xi_0)$$
$$(x_0 < \xi_0 < x_1),$$

provided that $f^{(2)}$ is continuous on $[x_0, x_1]$. Hence we have

$$\int_{x_0}^{x_1} f(x)\,dx = \frac{h}{2}[f_0 + f_1] - \frac{h^3}{12}f^{(2)}(\xi_0) \quad (x_0 < \xi_0 < x_1). \tag{3.40}$$

For integration over $[a, b]$ with notation as in Eqns (3.37), we have

$$\int_a^b f(x)\,dx = \frac{h}{2}[f_0 + 2f_1 + \ldots + 2f_{n-1} + f_n] - \frac{nh^3}{12}f^{(2)}(\xi)$$
$$(a < \xi < b). \tag{3.41}$$

*Example 3.7*

Consider Example 3.5 (2). With $f(x) = 1/x$, $a = 1$, $b = 2$ we have

$$|f^{(2)}(x)| = \left|\frac{2}{x^3}\right| \leqslant 2 \ (1 \leqslant x \leqslant 2).$$

With $h = 0\cdot 2$ the truncation error $E_T$ obtained using the trapezium rule to evaluate $\int_1^2 f(x)\,dx$ is bounded by

$$|E_T| \leqslant \frac{5 \times (0\cdot 2)^3 \times 2}{12} = 0\cdot 00\dot{6}.$$

Similarly with $h = 0\cdot 1$,

$$|E_T| \leqslant \frac{10 \times (0\cdot 1)^3 \times 2}{12} = 0\cdot 001\dot{6},$$

and with $h = 0\cdot 05$,

$$|E_T| = \frac{20 \times (0\cdot 05)^3 \times 2}{12} = 0\cdot 00041\dot{6}.$$

## §3.7] TRUNCATION ERROR IN INTERPOLATORY QUADRATURE

*Case* (3)

With $a = x_0$, $b = x_2$, and $n = 2$ in Eqn (3.32) we have for the truncation error $E_T$ corresponding to Simpson's rule

$$E_T = \frac{1}{6} \int_{x_0}^{x_2} \phi_2(x) f^{(3)}(\xi(x)) \, dx.$$

Now $\phi_2(x)$ changes sign at $x = x_1$.

However, we have

$$E_T = \tfrac{1}{6}[f^{(3)}(\eta_0) \int_{x_0}^{x_1} \phi_2(x) \, dx + f^{(3)}(\eta_1) \int_{x_1}^{x_2} \phi_2(x) \, dx]$$

where $x_0 < \eta_0 < x_1$, $x_1 < \eta_1 < x_2$, considering $[x_0, x_1]$ and $[x_1, x_2]$ separately. Now with $x = x_0 + sh$,

$$\int_{x_0}^{x_1} \phi_2(x) \, dx = h^4 \int_0^1 s(s-1)(s-2) \, ds = \frac{h^4}{4},$$

$$\int_{x_1}^{x_2} \phi_2(x) \, dx = h^4 \int_1^2 s(s-1)(s-2) \, ds = -\frac{h^4}{4},$$

so

$$E_T = \frac{h^4}{24}[f^{(3)}(\eta_0) - f^{(3)}(\eta_1)].$$

If $f^{(4)}$ exists on $(x_0, x_2)$ and $f^{(3)}$ is continuous on $[x_0, x_2]$ then by the mean value theorem

$$E_T = -\frac{h^4}{24}(\eta_1 - \eta_0) f^{(4)}(\xi) \quad (x_0 < \eta_0 < \xi < \eta_1 < x_2).$$

If we assume that $\eta_1 - \eta_0$ is independent of $f$ and set

$$\eta_1 - \eta_0 = Kh$$

where $K$ is a constant such that $0 < K \leqslant 2$, then we obtain

$$E_T = -\frac{Kh^5}{24} \cdot f^{(4)}(\xi) \quad (x_0 < \xi < x_2).$$

Then we have $\int_{x_0}^{x_2} f(x) \, dx = \frac{h}{3}[f_0 + 4f_1 + f_2] - \frac{Kh^5}{24} f^{(4)}(\xi).$

To obtain the value of $K$ we note that if $f(x) = x^4$ then $f^{(4)}(\xi) = 4!$, $\int_{x_0}^{x_2} f(x) \, dx = (2h)^5/5$, and $h[f_0 + 4f_1 + f_2]/3$

$= 20h^5/3$ so that $K = 4/15$, whence

$$\int_{x_0}^{x_2} f(x)\,\mathrm{d}x = \frac{h}{3}[f_0 + 4f_1 + f_2] - \frac{h^5}{90}f^{(4)}(\xi)$$
$$(x_0 < \xi < x_2). \qquad (3.42)$$

A lengthy argument shows that $(\eta_1 - \eta_0)$ is indeed independent of $f$ and Eqn (3.42) holds for all sufficiently differentiable functions. For an alternative elementary derivation of Eqn (3.42) see Massey and Kestelman (1959), or Scheid (1968).

For integration over $[a, b]$ we have

$$\int_a^b f(x)\,\mathrm{d}x = \frac{h}{3}[f_0 + 4f_1 + 2f_2 + \ldots + 4f_{2n-1} + f_{2n}]$$
$$- \frac{nh^5}{90}f^{(4)}(\xi) \qquad (3.43)$$

where $a < \xi < b$.

*Example 3.8*

Consider Example 3.5 (3). With $f(x) = 1/x$, $a = 1$, $b = 2$ we have

$$|f^{(4)}(x)| = \left|\frac{24}{x^5}\right| \leqslant 24 \quad (1 \leqslant x \leqslant 2).$$

With $h = 0 \cdot 2$ the truncation error $E_T$ obtained using Simpson's rule to evaluate $\int_1^2 f(x)\,\mathrm{d}x$ is bounded by

$$|E_T| \leqslant \frac{5 \times (0\cdot 2)^5 \times 24}{90} = 0\cdot 000426\dot{6}.$$

Similarly with $h = 0 \cdot 1$,

$$|E_T| \leqslant \frac{10 \times (0\cdot 1)^5 \times 24}{90} = 0\cdot 0000266\dot{6},$$

and with $h = 0 \cdot 05$,

$$|E_T| \leqslant \frac{20 \times (0\cdot 05)^5 \times 24}{90} = 0\cdot 00000016\dot{6}.$$

## 3.8 The Taylor Expansion Method for Numerical Integration

As in the case of numerical differentiation, any quadrature formula of specified type can be obtained, together with a bound on its truncation error, by using the Taylor expansion method. To illustrate the general method we shall obtain an integration formula of the form

$$I = \int_{x_0}^{x_1} f(x)\,dx = h(\alpha_1 f_0 + \alpha_2 f_1) + h^2(\beta_1 f_0^{(1)} + \beta_2 f_1^{(1)}) + E_T. \quad (3.44)$$

Let

$$F(t) = \int f(t)\,dt.$$

Then

$$F^{(1)}(t) = f(t), \quad F^{(k)}(t) = f^{(k-1)}(t) \quad (k = 2, 3, \ldots),$$

and

$$I = F(x_1) - F(x_0).$$

Hence by Taylor's theorem

$$I = F(x_1) - F(x_0) = hf_0 + \frac{h^2}{2}f_0^{(1)} + \frac{h^3}{6}f_0^{(2)} + \frac{h^4}{24}f_0^{(3)} + \frac{h^5}{120}f^{(4)}(\xi_1) \quad (x_0 < \xi_1 < x_1). \quad (3.45)$$

Also

$$f_1 = f_0 + hf_0^{(1)} + \frac{h^2}{2}f_0^{(2)} + \frac{h^3}{6}f_0^{(3)} + \frac{h^4}{24}f^{(4)}(\xi_2) \quad (x_0 < \xi_2 < x_1) \quad (3.46)$$

and

$$f_1^{(1)} = f_0^{(1)} + hf_0^{(2)} + \frac{h^2}{2}f_0^{(3)} + \frac{h^3}{6}f^{(4)}(\xi_3) \quad (x_0 < \xi_3 < x_1). \quad (3.47)$$

Using Eqns (3.45), (3.46), and (3.47) in Eqn (3.44) we have an identity for all sufficiently differentiable functions $f$. Equating coefficients of $f_0$, $f_0^{(1)}$, $f_0^{(2)}$, and $f_0^{(3)}$ on the right-hand side to those on the left, we obtain

$$\alpha_1 + \alpha_2 = 1$$
$$\alpha_2 + \beta_1 + \beta_2 = \tfrac{1}{2}$$
$$\frac{\alpha_2}{2} + \beta_2 = \tfrac{1}{6}$$
$$\frac{\alpha_2}{6} + \frac{\beta_2}{2} = \tfrac{1}{24},$$

whence $\alpha_1 = \alpha_2 = \tfrac{1}{2}$, $\beta_1 = -\beta_2 = \tfrac{1}{12}$, and

$$E_T = h^5 \left[ \frac{f^{(4)}(\xi_1)}{120} - \frac{f^{(4)}(\xi_2)}{48} + \frac{f^{(4)}(\xi_3)}{72} \right].$$

If

$$|f^{(4)}(x)| \leqslant M \quad (x_0 \leqslant x \leqslant x_1),$$

then we have the so-called *trapezoidal rule with end correction*,

$$\int_{x_0}^{x_1} f(x)\,dx = \frac{h}{2}[f_0 + f_1] + \frac{h^2}{12}[f_0^{(1)} - f_1^{(1)}] + E_T, \quad (3.48)$$

where

$$|E_T| \leqslant \tfrac{31}{720} M h^5.$$

It can be shown that there is a number $\xi$ in $(x_0, x_1)$ such that

$$E_T = \frac{h^5}{720} f^{(4)}(\xi). \tag{3.49}$$

For integration over $[a, b]$ using Eqn (3.48) we have, with the same notation as for the trapezoidal rule,

$$\int_a^b f(x)\,dx = \frac{h}{2}[f_0 + 2f_1 + \ldots + 2f_{n-1} + f_n]$$
$$+ \frac{h^2}{12}[f_0^{(1)} - f_n^{(1)}] + \frac{nh^5}{720} f^{(4)}(\xi)$$
$$(a < \xi < b). \tag{3.50}$$

*Exercise 3.1*

Using the data in Table 3.2 estimate the value of $\int_1^2 f(x)\,dx$ from Eqn (3.50). Obtain bounds on the truncation and rounding errors for this estimate. Note that $f^{(1)}(1)$ and $f^{(1)}(2)$ must be computed by differentiating $f$, and that rounding error in quadrature formulae is estimated in a similar manner to that for differentiation formulae.

### Tutorial Examples

1. Using Eqn (3.3) with $m = 2$ and $n = 4$, obtain the differentiation formula

$$f_2^{(1)} \approx \frac{1}{12h}[f_0 - 8f_1 + 8f_3 - f_4].$$

Use this formula to estimate the value of $\cos 0\cdot 302$ from the following table.

| $x$ | $\sin x$ |
|---|---|
| 0·300 | 0·295520 |
| 0·301 | 0·296475 |
| 0·302 | 0·297430 |
| 0·303 | 0·298385 |
| 0·304 | 0·299339 |

Compare the resulting estimate with that obtained using the formula

$$f_1^{(1)} \approx \frac{1}{2h}[f_2 - f_0],$$

with $h = 0\cdot 002$. Correct to 6D, $\cos 0\cdot 302 = 0\cdot 954744$.

2. By using Newton's backward difference formula obtain a differentiation formula of the form

$$f_0^{(1)} = \frac{1}{h}[\alpha_1 f_0 + \alpha_2 f_{-1} + \alpha_3 f_{-2}].$$

Use this formula to estimate the value of cos 0·304 using the data in Tutorial Example 3.1, firstly with $h = 0·001$ and secondly with $h = 0·002$. Compare the results with the exact value cos 0·304 = 0·954147 correct to 6D.

3. Obtain a bound for the truncation error resulting from the use of the first formula in Tutorial Example 3.1 and compare it with the corresponding bound for the second formula. How do you account qualitatively for the magnitude of the error in the estimate of cos 0·302 obtained using the first formula?

4. Obtain the first formula in Tutorial Example 3.1 and the formula in Tutorial Example 3.2 together with expressions for their truncation errors using the Taylor expansion method.

5. Obtain bounds for the rounding errors incurred by the formulae in Tutorial Examples 3.1 and 3.2 resulting from the rounding errors in the data. Hence account quantitatively for the errors obtained by using these formulae to estimate cos 0·302.

6. It is required to estimate $f^{(1)}(x)$ using the formula

$$f_{m+1}^{(1)} \approx \frac{1}{2h}[f_{m+2} - f_m]$$

from a table of values of $f$ which are correct to 6D. Given that

$$|f^{(3)}(x)| \leqslant \tfrac{3}{2}$$

on the interval of tabulation of $f$, what value of $h$ gives a maximum value for the bound on the total error due to truncation and rounding in the estimate of $f^{(1)}(x)$?

TUTORIAL EXAMPLES

7. Using the Taylor expansion method obtain a differentiation formula of the form

$$f_0^{(2)} \approx \frac{1}{h^2}[\alpha_1 f_0 + \alpha_2 f_1] + \frac{1}{h}[\beta_1 f_0^{(1)} + \beta_2 f_1^{(1)}].$$

Obtain bounds on its truncation and rounding errors. Use the formula to estimate $f^{(2)}(0·302)$ from the data in Tutorial Example 3.1 using the values of cos 0·302 and cos 0·304 given in Tutorial Examples 3.1 and 3.2. Compare the result with that obtained using the formula

$$f_1^{(2)} \approx \frac{1}{2h}[f_2^{(1)} - f_0^{(1)}]$$

given that cos 0·300 = 0·955336 correct to 6D. This example is another illustration of the sensitivity of numerical differentiation formulae to rounding error, and illustrates the disastrous effect of loss of significance.

8. The Taylor expansion method is one form of the so-called *method of undetermined coefficients*, in which a differentiation, integration or interpolation formula of given form is found by requiring, say, that it be exact for all polynomials of degrees $\leq n$ where $n$ is given. The resulting $(n + 1)$ conditions yield $(n + 1)$ linear equations for the undetermined coefficients in the formula.

Obtain $\alpha_1, \ldots, \alpha_4$ by requiring that the formula

$$f_2^{(1)} \approx \frac{1}{h}[\alpha_1 f_0 + \alpha_2 f_1 + \alpha_3 f_3 + \alpha_4 f_4]$$

be exact for the polynomials $1, x, x^2, x^3$.

9. Noting that

$$\int_0^1 \frac{dx}{(1 + x^2)} = \frac{\pi}{4},$$

estimate the value of $\pi$ from the following table using the rectangle rule, the trapezium rule, and Simpson's rule each with $h = 0·25$ and $h = 0·125$.

| $x$ | $1/(1 + x^2)$ |
|---|---|
| 0·000 | 1·0000000 |
| 0·125 | 0·9846154 |
| 0·250 | 0·9411765 |
| 0·375 | 0·8767123 |
| 0·500 | 0·8000000 |
| 0·625 | 0·7191011 |
| 0·750 | 0·6400000 |
| 0·875 | 0·5663717 |
| 1·000 | 0·5000000 |

10. Obtain bounds for the truncation errors in the estimates of $\pi$ found in Tutorial Example 3.9.

11. Using the Taylor expansion method obtain the midpoint rule,
$$\int_{x_0}^{x_2} f(x)\,dx = 2hf_1 + \frac{h^3}{3} f^{(2)}(\xi) \quad (x_0 < \xi < x_2).$$

    Use this rule with $h = 0.125$ to estimate the value of $\pi$ from the data in Tutorial Example 3.9, and obtain a bound for the truncation error in the estimate of $\pi$.

12. Obtain bounds for the rounding errors in the estimates of $\pi$ found in Tutorial Example 3.9.

13. Using the Taylor expansion method obtain Simpson's rule together with a bound on the truncation error. Obtain $\alpha_1$, $\alpha_2$, and $\alpha_3$ so that the formula
$$\int_{x_0}^{x_2} f(x)\,dx \approx h[\alpha_1 f_0 + \alpha_2 f_1 + \alpha_3 f_2]$$
is exact for $f(x) = 1$, $x$, and $x^2$. Note that it is then also exact for $f(x) = x^3$, since the truncation error in Simpson's rule depends upon $f^{(4)}(x)$.

CHAPTER 4

# Numerical Solution of Equations in One Real Variable

## 4.1 Introduction

A very important problem which arises frequently in applications of mathematics is that of finding numbers $x$ such that

$$f(x) = 0. \tag{4.1}$$

If $f$ is a polynomial then Eqn (4.1) is called a *polynomial equation* or an *algebraic equation*. If $f$ is not a polynomial then Eqn (4.1) is called a *transcendental equation*. The values $x^*$ of $x$, if they exist, which satisfy Eqn (4.1) are called the *roots* or *solutions* of Eqn (4.1), or the *zeros* of $f$.

In general there is no formula from which the roots of Eqn (4.1) can be obtained using a finite number of arithmetical operations, as there is, for example, when $f$ is a quadratic polynomial. In this chapter we shall consider some iterative methods for the numerical solution of Eqn (4.1), in which, starting with one or more initial estimates $x_0, x_1, \ldots$ of a root $x^*$, we compute a sequence $\{x_n\}$ of numbers which converges to $x^*$ as $n$ approaches infinity.

## 4.2 Existence of a Root

Given a function $f$ which is continuous on an interval containing $[a, b]$ then if $f(a)$ and $f(b)$ have opposite signs it is intuitively clear that there is at least one solution $x^*$ of Eqn (4.1) in $(a, b)$, because the graph of $f$ must cross the $x$-axis at least once between $x = a$ and $x = b$. Hence we have the following result, which is also easily proved analytically.

*Lemma 4.1*

If (1) $f$ is continuous on $[a, b]$;
   (2) $f(a)f(b) < 0$,

then Eqn (4.1) has at least one root in $(a, b)$. ∎

From Lemma 4.1 we obtain another useful result.

*Lemma 4.2*

If (1) $F$ is continuous on $[a, b]$;
   (2) $a \leqslant x \leqslant b$ implies that $a \leqslant F(x) \leqslant b$,

then there is at least one number $x^*$ in $[a, b]$ such that

$$x^* = F(x^*).$$

*Proof*

Let

$$G(x) = x - F(x).$$

Then by hypothesis (2),

$$G(a) \leqslant 0, \qquad G(b) \geqslant 0.$$

Hence by hypothesis (1) and Lemma 4.1 there is at least one number $x^*$ in $[a, b]$ such that

$$G(x^*) = 0. \blacksquare$$

## 4.3 Uniqueness of a Root

To establish sufficient conditions for the root $x^*$ of the equation

$$x = F(x) \tag{4.2}$$

to be unique in a given interval we introduce the idea of a *Lipschitz condition*.

Now if $F^{(1)}$ is continuous on $[a, b]$ then given any two numbers $x'$ and $x''$ in $[a, b]$, we have by the mean value theorem

$$|F(x') - F(x'')| \leqslant L|x' - x''|, \tag{4.3}$$

where

$$|F^{(1)}(x)| \leqslant L \quad (a \leqslant x \leqslant b).$$

The inequality (4.3) is called a *Lipschitz condition*, and $L$ is called a *Lipschitz constant*.

If $F$ satisfies a Lipschitz condition on $[a, b]$ then $F$ is continuous on $[a, b]$, for by hypothesis, if $x'$ and $x''$ are any two points on $[a, b]$ then as $x' \to x''$, $F(x') \to F(x'')$. We conclude from this that a function which is not continuous on $[a, b]$ cannot satisfy a Lipschitz condition on $[a, b]$. However $F$ need not be differentiable on $[a, b]$ in order to satisfy a Lipschitz condition on $[a, b]$.

We are now able to establish the following important result.

*Lemma 4.3*

If (1) $a \leqslant x \leqslant b$ implies that $a \leqslant F(x) \leqslant b$;
 (2) $|F(x') - F(x'')| \leqslant L|x' - x''|$
 $\qquad\qquad\qquad\qquad (a \leqslant x' \leqslant b, a \leqslant x'' \leqslant b)$;
 (3) $0 \leqslant L < 1$,

then there exists one and only one root $x^*$ of Eqn (4.2) in $[a, b]$.

*Proof*

Noting that hypothesis (2) implies the continuity of $F$ on $[a, b]$ we conclude from Lemma 4.2 that there is at least one root $x^*$ of Eqn (4.2) in $[a, b]$. Assume that there are two distinct roots $x^*$ and $x^{**}$ in $[a, b]$. Then

$$|F(x^*) - F(x^{**})| \leqslant L|x^* - x^{**}|,$$
$$F(x^*) = x^*, \quad F(x^{**}) = x^{**}.$$

So

$$|x^* - x^{**}| \leqslant L|x^* - x^{**}|.$$

Since $x^* \neq x^{**}$ we can divide by $|x^* - x^{**}|$ whence $1 \leqslant L$, contrary to hypothesis (3). Hence if hypothesis (3) holds, then $x^{**} = x^*$, and so $x^*$ is unique in $[a, b]$. ∎

*Example 4.1*

Consider the equation
$$x = \cos x,$$
and let $a = \pi/6$, $b = \pi/4$ so that with
$$F(x) = \cos x,$$
$F(x)$ is strictly decreasing on $[a, b]$ and
$$F(a) = \frac{\sqrt{3}}{2}, \quad F(b) = \frac{1}{\sqrt{2}}.$$

Also,
$$|F^{(1)}(x)| = |\sin x| \leqslant \frac{1}{\sqrt{2}} = L < 1 \quad \left(\frac{\pi}{6} \leqslant x \leqslant \frac{\pi}{4}\right).$$

Hence $F$ satisfies the hypotheses of Lemma 4.3 on $[\pi/6, \pi/4]$ and so there is one and only one root of the given equation in this interval.

*Example 4.2*

Consider the equation
$$2x = \exp(-x)$$
and let $a = 0$, $b = 1$, so that with
$$F(x) = \frac{\exp(-x)}{2},$$
$F(x)$ is strictly decreasing on $[a, b]$ and
$$F(a) = 1/2, \quad F(b) = 1/2e.$$

Also,
$$|F^{(1)}(x)| = \left|\frac{\exp(-x)}{2}\right| \leqslant 1/2 = L < 1 \quad (0 \leqslant x \leqslant 1).$$

Hence $F$ satisfies the hypothesis of Lemma 4.3 on $[0, 1]$ and so there is one and only one root of the given equation in this interval.

## 4.4 The Method of Bisection

Probably the simplest method for the numerical solution of Eqn (4.1) is the method of bisection, which is illustrated in Fig. 4.1. Let $f$ be a continuous function of $x$ on $[a, b]$, the graph of which crosses the $x$-axis at $x = x^*$ such that $a \leqslant x_0 < x^* < x_1 \leqslant b$.

Then $f(x_0)$ and $f(x_1)$ have opposite signs. Let

$$x_2 = \frac{(x_0 + x_1)}{2}.$$

Then $x^*$ lies either in $(x_0, x_2)$ or in $(x_2, x_1)$, or $x^* = x_2$ in which case $x^*$ is found. Suppose that $x_2 < x^* < x_1$. Let

$$x_3 = \frac{(x_1 + x_2)}{2}.$$

Then if $x^* \neq x_3$, $x^*$ lies either in $(x_2, x_3)$ or in $(x_3, x_1)$. Clearly this bisection procedure could be repeated indefinitely with $x^*$ known to be in a sequence of intervals each of which has a length of half that of the preceding interval. A simple criterion that $x^*$ lies in $(x_i, x_j)$ is that $f(x_i)$ and $f(x_j)$ have opposite signs, in which case

$$f(x_i)f(x_j) < 0.$$

Let $\delta_n$ ($n = 0, 1, 2, \ldots$) be the lengths of the intervals obtained by bisection and containing $x^*$ so that for the case illustrated in Fig. 4.1,

$$\begin{aligned}
\delta_0 &= |x_1 - x_0|, \\
\delta_1 &= |x_1 - x_2| = \delta_0/2, \\
\delta_2 &= |x_1 - x_3| = \delta_1/2 = \delta_0/2^2, \\
&\ldots \\
\delta_n &= \delta_0/2^n.
\end{aligned}$$

Hence the length of the interval containing $x^*$ after $n$ bisections is $\delta_n$, and if the final estimate $\hat{x}$ of $x^*$ is taken to be the midpoint of this interval, then

$$e_n = |x^* - \hat{x}| \leqslant \delta_0/2^{n+1}. \tag{4.4}$$

So as $n \to \infty$, $\hat{x} \to x^*$.

The number of evaluations of $f$ needed for $n$ bisections is $(n+2)$ counting the two preliminary evaluations at $x = x_0$ and $x = x_1$. The accuracy with which $x^*$ can be estimated is limited only by the accuracy with which the sign of $f(x)$ can be determined. The method of bisection makes no use of the

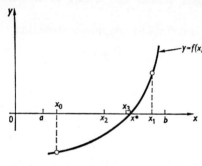

Fig. 4.1

properties of $f$ and so will always work, although the convergence is clearly slow.

We can express the method of bisection in the form of an algorithm as follows.

*Algorithm 4.1*

1. Compute $f(x_0)$ and $f(x_1)$ and go to 2.
2. Compute $x_2$ from

$$x_2 = \frac{(x_0 + x_1)}{2},$$

and if $|x_1 - x_2| \leq \varepsilon$, set $\hat{x} = x_2$ and stop; otherwise go to 3.

3. Compute $f(x_2)$ and go to 4.
4. If $f(x_2)f(x_0) < 0$ set $x_1 = x_2$ and go to 2; otherwise set $x_0 = x_2$ and go to 2. ■

*Example 4.3*

The equation

$$f(x) = 8x^3 + 8x - 5 = 0$$

is known to have a root $x^*$ in (0·3, 0·6) at which the graph of $f$ crosses the $x$-axis. We shall estimate the value of $x^*$ correct to 2D using Algorithm 4.1. Table 4.1 shows the values of $x_0$, $x_1$, and $x_2$ at each stage of implementation of the algorithm. Clearly from inequality (4.4) we must take $\varepsilon = 0·005$ if we require an accuracy of 2D. From Table 4.1 we obtain $x^* = 0·50$ correct to 2D. The exact value of $x^*$ is $\frac{1}{2}$.

| $x_0$ | $x_1$ | $x_2$ | $|x_1 - x_2|$ | $f(x_0)f(x_2)$ |
|---|---|---|---|---|
| 0·300 | 0·600 | 0·450 | 0·150 | >0 |
| 0·450 | 0·600 | 0·525 | 0·075 | <0 |
| 0·450 | 0·525 | 0·488 | 0·037 | >0 |
| 0·488 | 0·525 | 0·506 | 0·019 | <0 |
| 0·488 | 0·506 | 0·497 | 0·009 | >0 |
| 0·497 | 0·506 | 0·502 | 0·004 | |

TABLE 4.1

The number $n$ of bisections which ensure an accuracy of $m$D is easily obtained from inequality (4.4) by choosing $n$ such that

$$|x_1 - x_0|10^m \leqslant 2^n.$$

For Example 4.3, this gives $2^n \geqslant 30$ whence $n = 6$ as we have seen.

## 4.5 The Method of Iteration

To solve Eqn (4.1) numerically by the method of iteration, we express Eqn (4.1) in the form of Eqn (4.2), and with $x_0$ given, we generate the sequence of *iterates* $\{x_n\}$ from

$$x_{n+1} = F(x_n) \quad (n = 0, 1, 2, \ldots). \tag{4.5}$$

The function $F$, which clearly depends upon $f$, is called an *iteration function*. If Eqn (4.1) has a root $x^*$ and if $F$ and $x_0$ are properly chosen then $x_n \to x^*$ as $n \to \infty$. We have the following theorem.

## THEOREM 4.1

If (1) the hypotheses of Lemma 4.3 are valid;
 (2) $a \leqslant x_0 \leqslant b$,

then there exists a unique solution $x^*$ of Eqn (4.2) in $[a, b]$ to which the sequence $\{x_n\}$ generated from Eqn (4.5) converges.

*Proof*

By hypotheses (1), (2), and (3) of Lemma 4.3 there exists a unique root $x^*$ of Eqn (4.2) in $[a, b]$. By hypotheses (1) of Lemma 4.3 and (2) of this theorem, $a \leqslant x_n \leqslant b$ for all values of $n \geqslant 0$. Hence by hypothesis (2) of Lemma 4.3

$$|F(x^*) - F(x_{n-1})| = |x^* - x_n| \leqslant L|x^* - x_{n-1}| \quad (n = 1, 2, \ldots).$$

Hence by repeated use of this inequality,

$$|x^* - x_n| \leqslant L^n |x^* - x_0| \quad (n = 0, 1, 2, \ldots).$$

So as $n \to \infty$, $|x^* - x_n| \to 0$ by hypothesis (3) of Lemma 4.3 so $x_n \to x^*$ as $n \to \infty$. ∎

*Example 4.4*

By drawing a graph or by evaluation of $f$ at a few points it is easily established that there is a root $x^*$ of

$$f(x) = x^2 \exp x + 2x - 1 = 0$$

in $(0.3, 0.5)$, and that $x^*$ is near $0.4$. Equation (4.1) can be expressed in the form of Eqn (4.2) by taking

$$F(x) = x - \alpha f(x) \quad (a \leqslant x \leqslant b)$$

where $\alpha$ is any finite number. Then Eqns (4.1) and (4.2) have the same roots. Take $\alpha = 1/4$. Then $F(x)$ is strictly increasing on $[0.3, 0.5]$ with $F(0.3) > 0.3$ and $F(0.5) < 0.5$. Also

$$|F^{(1)}(x)| = \left| \frac{1}{2} - \left( \frac{x}{2} + \frac{x^2}{4} \right) \exp x \right| < 0.27 = L < 1$$

$$(0.3 \leqslant x \leqslant 0.5).$$

Hence with $x_0 = 0.4$, the hypotheses of Theorem 4.1 are valid on [0·3, 0·5], so there is a unique root $x^*$ of the given equation in [0·3, 0·5] to which the sequence $\{x_n\}$ generated from

$$x_{n+1} = x_n - \frac{(x_n^2 \exp x_n + 2x_n - 1)}{4} \quad (n = 0, 1, 2, \ldots)$$

converges. With

$$e_n = x^* - x_n \quad (n = 0, 1, 2, \ldots) \tag{4.6}$$

the numerical results are shown in Table 4.2.

| $n$ | $x_n$ | $e_n$ | $e_{n+1}/e_n$ |
|---|---|---|---|
| 0 | 0·400000 | 0·011378 | 0·149851 |
| 1 | 0·390327 | 0·001705 | 0·156598 |
| 2 | 0·388890 | 0·000267 | 0·157303 |
| 3 | 0·388664 | 0·000042 | 0·166667 |
| 4 | 0·388629 | 0·000007 | |
| 5 | 0·388623 | 0·000001 | |
| 6 | 0·388622 | 0·000000 | |

TABLE 4.2

In Table 4.2, the values of $e_n$ after $n = 3$ are insufficiently accurate to indicate that $e_{n+1}/e_n$ approaches constancy as $n$ increases.

The method of iteration may be expressed algorithmically as follows.

*Algorithm 4.2*

1. Compute $x_1$ from

$$x_1 = F(x_0),$$

and go to 2.

2. If $x_1$ satisfies a given convergence criterion, set $x^* = x_1$ and stop; otherwise go to 3.

3. Set $x_0 = x_1$ and go to 1. ∎

Especially when implementing Algorithm 4.2 on a computer it is necessary to have a criterion for terminating the iterative sequence, and to allow for the possible non-convergence of the sequence by causing the iteration to stop after a pre-assigned number of iterates have been computed. Possible convergence criteria are as follows.

(a) If $|x_1 - x_0| \leqslant \varepsilon$, stop.
(b) If $|x_1 - x_0| \leqslant \lambda|x_1|$, stop.
(c) If $|f(x_1)| \leqslant \delta$, stop.

In these criteria, $\varepsilon$, $\lambda$, and $\delta$ are given numbers. Criterion (a) does not necessarily ensure that the absolute error in $x_1$ is $\varepsilon$, but is useful when $x^*$ is of order unity. Criterion (b) does not always ensure a relative error $\lambda$ in $x_1$ but is useful when $x^* \ll 1$ or $x^* \gg 1$. If the rate of convergence of $\{x_n\}$ is slow then criterion (a) or criterion (b) may be satisfied by a value $\hat{x}$ of $x$ even though $f(\hat{x})$ is far from zero. In doubtful cases it may be worth while to use criteria (a) and (c) together or criteria (b) and (c) together.

We can calculate the number of iterations which are required to obtain a given accuracy when using Algorithm 4.2 using the following theorem.

*THEOREM 4.2*

If the hypotheses of Theorem 4.1 are valid, then

$$|x^* - x_n| \leqslant \frac{L^n}{(1-L)}|x_1 - x_0|.$$

*Proof*

Since $a \leqslant x_n \leqslant b$ for all $n \geqslant 0$, we have

$$|x_n - x_{n+1}| = |F(x_{n-1}) - F(x_n)| \leqslant L|x_{n-1} - x_n|$$
$$(n = 1, 2, \ldots).$$

By repeated use of this inequality we obtain

$$|x_n - x_{n+1}| \leqslant L^n|x_0 - x_1| \quad (n = 0, 1, 2, \ldots).$$

Hence for any positive integer $m$,

$$|x_n - x_{n+m}| = |x_n - x_{n+1} + x_{n+1} - x_{n+2} + \ldots + x_{n+m-1} - x_{n+m}|$$
$$\leqslant |x_n - x_{n+1}| + \ldots + |x_{n+m-1} - x_{n+m}|$$
$$\leqslant L^n|x_0 - x_1|(1 + L + L^2 + \ldots + L^{m-1})$$
$$(n = 0, 1, 2, \ldots).$$

Keeping $n$ fixed and letting $m \to \infty$, we have, since $x_{n+m} \to x^*$ as $m \to \infty$, and on summing the infinite geometrical progression,

$$|x_n - x^*| \leqslant \frac{L^n}{(1-L)}|x_0 - x_1| \quad (n = 0, 1, 2, \ldots).\blacksquare$$

To obtain an accuracy of $m$D we then need $n$ iterations, where

$$\frac{L^n}{(1-L)}|x_0 - x_1| \leqslant \frac{10^{-m}}{2}.$$

*Example 4.5*

In Example 4.4, with $L = 0.27$, $|x_0 - x_1| = 0.009678$, and $m = 5$, we obtain $n \geqslant 7$ so 7 iterations will ensure 5 correct decimal places. In fact only 5 iterations are needed as seen from Table 4.2.

When it is known that Eqn (4.2) has a root $x^*$ and that $F^{(1)}(x)$ is bounded in a neighbourhood of $x^*$ the following theorem is easier to apply than Theorem 4.1.

*THEOREM 4.3*

If (1) Eqn (4.2) has a root $x^*$;
  (2) $|F^{(1)}(x)| \leqslant L < 1 \quad (x^* - \rho \leqslant x \leqslant x^* + \rho)$;
  (3) $x^* - \rho \leqslant x_0 \leqslant x^* + \rho$,

then the root $x^*$ is unique in $[x^* - \rho, x^* + \rho]$ and the sequence $\{x_n\}$ generated from Eqn (4.5) converges to $x^*$.

*Proof*

Firstly we show by induction that $x^* - \rho \leqslant x_n \leqslant x^* + \rho$ for all $n \geqslant 0$. Assume it true for some $n \geqslant 0$. Then by the mean value theorem,

$$|F(x^*) - F(x_n)| = |x^* - x_{n+1}| = |F^{(1)}(\xi_n)| \, |x^* - x_n|,$$

where $\xi_n$ lies between $x_n$ and $x^*$. Hence by hypothesis (2),

$$|x^* - x_{n+1}| \leqslant L|x^* - x_n| \leqslant L\rho < \rho.$$

Hence $x^* - \rho \leqslant x_{n+1} \leqslant x^* + \rho$. Hence by induction, in view of hypothesis (3), $x^* - \rho \leqslant x_n \leqslant x^* + \rho$ for all $n \geqslant 0$. Also by repeated use of the preceding inequality, we see that $x_n \to x^*$ as $n \to \infty$. Uniqueness is proved as in Lemma 4.3. ∎

*Example 4.6*

By using tables or by drawing a graph it is easily established that the equation

$$x = (\sin x)^{1/2} = F(x)$$

has a root $x^* = 0\cdot 88$ correct to 2D. Also, on $[\pi/6, \pi/2]$, $|F^{(1)}(x)| < 1$. Hence with $x_0 = 0\cdot 88$ the hypotheses of Theorem 4.3 are valid so that the sequence $\{x_n\}$ generated from

$$x_{n+1} = (\sin x_n)^{1/2} \quad (n = 0, 1, 2, \ldots)$$

will converge to $x^*$.

Note that hypothesis (2) of Theorem 4.3 ensures that $F$ satisfies a Lipschitz condition on $[x^* - \rho, x^* + \rho]$ and since we have shown that $x^* - \rho \leqslant x_n \leqslant x^* + \rho$ for all $n \geqslant 0$, then Theorem 4.2 is also valid under the hypotheses of Theorem 4.3.

## 4.6 Linear Convergence

From Theorem 4.2 we conclude that if $L$ is close to unity the sequence $\{x_n\}$ converges very slowly. To obtain more rapidly convergent iterative sequences we must study more closely the manner in which $e_n$ defined by Eqn (4.6) decreases as $n$ increases.

If the hypotheses of Theorem 4.3 are valid we have seen that

$$e_{n+1} = e_n F^{(1)}(\xi_n) \quad (n = 0, 1, 2, \ldots)$$

where $\xi_n$ lies between $x_n$ and $x^*$. Since $x_n \to x^*$ as $n \to \infty$, we obtain as $n \to \infty$,

$$\frac{e_{n+1}}{e_n} \to F^{(1)}(x^*). \tag{4.7}$$

Hence

$$e_{n+1} = [F^{(1)}(x^*) + \varepsilon_n] e_n \quad (n = 0, 1, 2, \ldots), \tag{4.8}$$

where $\varepsilon_n \to 0$ as $n \to \infty$.

## Definition 4.1

Given any sequence of numbers $\{x_n\}$ convergent to $x^*$ in such a way that

$$e_{n+1} = (K + \varepsilon_n)e_n \quad (n = 0, 1, 2, \ldots)$$

where $\varepsilon_n \to 0$ as $n \to \infty$, and $K$ is a constant independent of $n$ such that $0 < |K| < 1$, the sequence $\{x_n\}$ is said to converge *linearly*. If

$$e_{n+1} = Ke_n \quad (n = 0, 1, 2, \ldots),$$

then $\{x_n\}$ is said to converge *geometrically*. ■

We see that the sequence $\{x_n\}$ corresponding to Eqn (4.7) converges linearly with $K = F^{(1)}(x^*)$, if $F^{(1)}(x^*) \neq 0$.

## 4.7 Aitken's $\Delta^2$ Process

We now consider a method for accelerating the convergence of any linearly convergent sequence of numbers no matter how they were generated. If the sequence $\{x_n\}$ is linearly convergent to $x^*$, then with the notation of Definition 4.1, for any $n \geq 0$,

$$x_{n+1} - x^* = K(x_n - x^*) + \varepsilon_n(x_n - x^*),$$
$$x_{n+2} - x^* = K(x_{n+1} - x^*) + \varepsilon_{n+1}(x_{n+1} - x^*).$$

Neglecting the second terms on the right-hand sides we obtain

$$x_{n+1} - x^* \approx K(x_n - x^*), \quad x_{n+2} - x^* \approx K(x_{n+1} - x^*).$$

Eliminating $K$, we obtain

$$x^* \approx x_n - \frac{(x_{n+1} - x_n)^2}{(x_{n+2} - 2x_{n+1} + x_n)}.$$

This strongly suggests that the sequence $\{\hat{x}_n\}$ defined by

$$\hat{x}_n = x_n - \frac{(x_{n+1} - x_n)^2}{(x_{n+2} - 2x_{n+1} + x_n)} \quad (n = 0, 1, 2, \ldots) \quad (4.9)$$

may converge more rapidly than $\{x_n\}$ to $x^*$, provided that $\hat{x}_n$ is defined for all $n \geq 0$. In fact we have the following theorem.

## THEOREM 4.4

If $\{x_n\}$ is a sequence convergent linearly to $x^*$ then the sequence $\{\hat{x}_n\}$ defined by Eqns (4.9) converges to $x^*$ more rapidly in the sense that

$$\hat{e}_n/e_n \to 0 \quad \text{as} \quad n \to \infty,$$

where

$$\hat{e}_n = x^* - \hat{x}_n \quad (n = 0, 1, 2, \ldots). \blacksquare$$

The procedure whereby $\{\hat{x}_n\}$ is generated from $\{x_n\}$ as explained is called *Aitken's $\Delta^2$ process* because of the form of the right-hand side of Eqn (4.9).

If Eqn (4.9) is applied to the first three iterates $x_0$, $x_1$, $x_2$ of a linearly convergent sequence generated from Eqn (4.5), then $\hat{x}_0$ may well be a better approximation than $x_0$, $x_1$, or $x_2$ to the root $x^*$ of Eqn (4.2). Regarding $\hat{x}_0$ as a new initial iterate $x_0$ we compute $x_1$ and $x_2$ using Eqn (4.5). If $\hat{x}_0$ is closer to $x^*$ than is $x_0$, we expect the new sequence generated from Eqn (4.5) with $x_0 = \hat{x}_0$ to be linearly convergent to $x^*$ also. It then seems reasonable to apply Eqn (4.9) to the first three iterates of the new sequence, and so on. The following algorithm for the numerical solution of Eqn (4.2) is suggested.

*Algorithm 4.3*

1. Compute $x_1$ from

$$x_1 = F(x_0),$$

and go to 2.

2. If $x_1$ satisfies a given convergence criterion, set $x^* = x_1$ and stop; otherwise go to 3.

3. Compute $x_2$ from

$$x_2 = F(x_1),$$

and go to 4.

4. If $x_2$ satisfies a given convergence criterion set $x^* = x_2$ and stop; otherwise go to 5.

5. Compute $\hat{x}_0$ from Eqn (4.9) with $n = 0$, and go to 6.

6. If $x_0$ satisfies a given convergence criterion set $x^* = \hat{x}_0$ and stop; otherwise go to 7.

7. Set $x_0 = \hat{x}_0$ and go to 1. ■

The procedure described in Algorithm 4.3 is essentially Steffensen iteration† as distinct from Aitken's $\Delta^2$ process.

*Example 4.7*

It is easily seen that the sequence $\{x_n\}$ generated in Example 4.4 and shown in Table 4.2 is linearly convergent. Table 4.3 shows the result of applying Algorithm 4.3 to the equation in Example 4.4, the letter A denoting the result of applying Eqn (4.9) to the preceding three iterates. Comparing Tables 4.2 and 4.3 we see that two evaluations of $f$ have been saved.

| $n$ | $x_n$ |
|---|---|
| 0 | 0·400000 |
| 1 | 0·390327 |
| 2 | 0·388889 |
| A | 0·388638 |
| 1 | 0·388624 |
| 2 | 0·388622 |
| A | 0·388622 |

TABLE 4.3

## 4.8 Superlinear Convergence

To obtain more rapidly convergent iterative sequences it is desirable to construct iteration functions $F$ which give rise to convergent sequences for which as $n \to \infty$,

$$\frac{|e_{n+1}|}{|e_n|^p} \to \kappa$$

where $e_n$ ($n = 0, 1, 2, \ldots$) is defined by Eqn (4.6), and $p > 1$. Convergence of this type is called *superlinear convergence*, $p$ is

† See Henrici (1964).

called the *order* of convergence, and $\kappa$ is called the *asymptotic error constant*.

Suppose Eqn (4.2) has a root $x^*$, that $F^{(p)}$ is continuous on a neighbourhood of $x^*$ containing all the iterates $x_n$ ($n = 0, 1, 2, \ldots$), and that $x_n \to x^*$ as $n \to \infty$. Then by Taylor's theorem, for $n = 0, 1, 2, \ldots$,

$$x_{n+1} = F(x_n) = F(x^* + e_n)$$
$$= F(x^*) + \sum_{k=1}^{p-1} \frac{e_n^k}{k!} F^{(k)}(x^*) + \frac{e_n^p}{p!} F^{(p)}(\xi_n), \quad (4.10)$$

where $\xi_n$ lies between $x_n$ and $x^*$. Hence if

$$F^{(k)}(x^*) = 0 \quad (k = 1, \ldots, p-1), \quad F^{(p)}(x^*) \neq 0, \quad (4.11)$$

then from Eqn (4.10), we have, as $n \to \infty$,

$$\frac{e_{n+1}}{e_n^p} \to \frac{F^{(p)}(x^*)}{p!}. \quad (4.12)$$

Hence if conditions (4.11) are satisfied, the order of convergence is $p$. We then say that $F$ is *of order p*.

If there is a number $\rho$ such that

$$\left| \frac{F^{(p)}(x)}{p!} \right| \leqslant L \quad (x^* - \rho \leqslant x \leqslant x^* + \rho), \quad (4.13)$$

and

$$L\rho^{p-1} < 1, \quad (4.14)$$

then with $x_0$ in $[x^* - \rho, x^* + \rho]$, we can show that $x_n$ is in $[x^* - \rho, x^* + \rho]$ for all $n \geqslant 0$ by induction. For by Eqns (4.10) and (4.11), assuming that $x_m$ is in $[x^* - \rho, x^* + \rho]$ for $m = 0, \ldots, n$,

$$|e_{n+1}| = |e_n|^p \left| \frac{F^{(p)}(\xi_n)}{p!} \right| \leqslant L|e_n|^p \leqslant L\rho^p < \rho,$$

whence $x_{n+1}$ is in $[x^* - \rho, x^* + \rho]$. Hence $x_n$ is in $[x^* - \rho, x^* + \rho]$ for all $n \geqslant 0$. Then

$$|e_{n+1}| \leqslant L|e_n|^p = L|e_n|^{p-1}|e_n|$$
$$\leqslant L\rho^{p-1}|e_n| \quad (n = 0, 1, 2, \ldots).$$

By repeated application of this inequality we obtain

$$|e_{n+1}| \leqslant (L\rho^{p-1})^{n+1}|e_0| \quad (n = 0, 1, 2, \ldots).$$

Hence as $n \to \infty$, $|e_n| \to 0$, and $x_n \to x^*$. We have established the following theorem.

## THEOREM 4.5

If (1) Eqn (4.2) has a root $x^*$;
(2) Conditions (4.11) hold;
(3) $F^{(p)}(x)$ is continuous and inequality (4.13) is valid;
(4) inequality (4.14) is valid;
(5) $x^* - \rho \leqslant x_0 \leqslant x^* + \rho$,

then the sequence $\{x_n\}$ generated from Eqn (4.5) converges to $x^*$. ∎

We illustrate this theorem in the following section.

## 4.9 Newton's Method

Clearly, by making $\rho$ sufficiently small, we can *always* ensure that hypothesis (4) of Theorem 4.5 is valid if $p > 1$, and $\{x_n\}$ then converges superlinearly. If however $p = 1$, then unless $L < 1$, $\{x_n\}$ will not converge to $x^*$ no matter how close to $x^*$ we choose $x_0$. Hence if the order $p$ of $F$ is greater than unity we can always generate a sequence $\{x_n\}$ convergent to a root $x^*$ of Eqn (4.2) by choosing $x_0$ sufficiently close to $x^*$, and this sequence will converge to $x^*$ more rapidly than any linearly convergent sequence. It is found that second-order or *quadratic* convergence is adequate for most purposes. Hence we seek a second-order iteration function which satisfies the hypotheses of Theorem 4.5.

Consider

$$F(x) = x - \frac{f(x)}{f^{(1)}(x)}, \tag{4.15}$$

which is defined on an interval throughout which $f^{(1)}(x)$ is non-zero. Clearly with this iteration function, Eqn (4.2) has the same roots as Eqn (4.1). Also, assuming the existence of $f^{(2)}$ and $f^{(3)}$,

$$F^{(1)} = \frac{f f^{(2)}}{[f^{(1)}]^2},$$

and

$$F^{(2)} = \frac{[f^{(1)}]^2 f^{(2)} + f[f^{(1)}]^2 f^{(3)} - 2f[f^{(2)}]^2}{[f^{(1)}]^3}.$$

Hence if Eqn (4.1) has a root $x^*$, then

$$F^{(1)}(x^*) = 0, \quad F^{(2)}(x^*) = \frac{f^{(2)}(x^*)}{f^{(1)}(x^*)}.$$

So if $f^{(1)}(x^*)$ and $f^{(2)}(x^*)$ are non-zero, $F^{(2)}(x^*)$ is non-zero and Theorem 4.5 ensures that if $x_0$ is chosen sufficiently closely to $x^*$, the sequence $\{x_n\}$ generated from

$$x_{n+1} = x_n - \frac{f(x_n)}{f^{(1)}(x_n)} \quad (n = 0, 1, 2, \ldots) \qquad (4.16)$$

converges quadratically to $x^*$.

This is Newton's method for the numerical solution of Eqn (4.1).

*Example 4.8*

The equation

$$f(x) = x^2 - \alpha = 0,$$

where $\alpha$ is any positive real number, has the root $x^* = \alpha^{1/2}$, and

$$f^{(1)}(x^*) = 2x^*, \quad f^{(2)}(x^*) = 2, \quad f^{(3)}(x^*) = 0.$$

Hence there is a number $\rho$ such that the hypotheses of Theorem 4.4 are satisfied and so if we choose $x_0$ sufficiently close to $\alpha^{1/2}$, the sequence $\{x_n\}$ generated from

$$x_{n+1} = F(x_n) = \frac{1}{2}\left(x_n + \frac{\alpha}{x_n}\right) \quad (n = 0, 1, 2, \ldots)$$

will converge quadratically to $\alpha^{1/2}$. In fact it can be shown†

† See Tutorial Example 4.7.

that $\{x_n\}$ converges to $\alpha^{1/2}$ for *any* choice of the positive real number $x_0$.

## 4.10 The Secant Method

Newton's method has a simple geometrical interpretation which is illustrated in Fig. 4.2. If we approximate the graph of

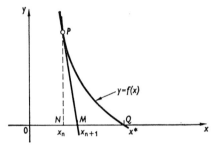

Fig 4.2

$f$ between $P$ and $Q$ by the tangent $PM$ then $x_{n+1}$ may be regarded as a new estimate of $x^*$. From Fig. 4.2,

$$f^{(1)}(x_n) = -\frac{PN}{NM} = -\frac{f(x_n)}{(x_{n+1} - x_n)},$$

from which we recover Eqn (4.16).

This geometrical approach suggests a method for approximating $f^{(1)}(x_n)$ which is useful when $f^{(1)}(x)$ is difficult to calculate analytically or to compute numerically, and which is illustrated in Fig. 4.3. The tangent to the graph of $f$ at $R$ is approximated by the secant $PRT$. Then

$$\frac{PN}{NT} = \frac{RS}{ST},$$

so

$$\frac{f(x_{n-1})}{(x_{n+1} - x_{n-1})} = \frac{f(x_n)}{(x_{n+1} - x_n)},$$

whence

$$x_{n+1} = x_n - \frac{[x_n - x_{n-1}]f(x_n)}{[f(x_n) - f(x_{n-1})]}. \quad (4.17)$$

Equation (4.17) suggests a new iterative method for the numerical solution of Eqn (4.1) called the *secant method*, which is contained in the following algorithm.

*Algorithm 4.4*

Given two estimates $x_0$ and $x_1$ of a root $x^*$ of Eqn (4.1).
1. Compute $x_2$ from Eqn (4.17) with $n = 1$ and go to 2.

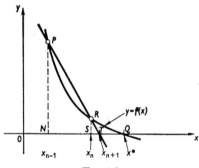

FIG 4.3

2. If $x_2$ satisfies a given convergence criterion set $x^* = x_2$ and stop; otherwise go to 3.

3. Set $x_0 = x_1$, $x_1 = x_2$ and go to 1. ∎

Note that only *one* evaluation of $f$ per iteration is needed. The secant method may therefore be more computationally efficient than Newton's method, even though the order of convergence of the secant method lies between 1 and 2.

*Example 4.9*

The equation

$$f(x) = x - \cos x = 0$$

has a root $x^*$ near 0·74. Table 4.4 shows the result of applying Algorithm 4.4 with $x_0 = 0·72$, $x_1 = 0·73$. If we take

$$F(x) = \cos x,$$

and use Algorithm 4.2 with $x_0 = 0.74$, *twenty-one* iterations are needed to estimate $x^*$ correct to 6D. Table 4.4 shows that using Algorithm 4.4 only *five* iterations are needed to estimate $x^*$ correct to 8D!

| $n$ | $x_n$ |
|---|---|
| 0 | 0·72000000 |
| 1 | 0·73000000 |
| 2 | 0·73912406 |
| 3 | 0·73908508 |
| 4 | 0·73908514 |
| 5 | 0·73908514 |

TABLE 4.4

## 4.11 The Method of False Position

Suppose that the graph of $f$ cuts the x-axis once in $(x_0, x_1)$. Given the values $f(x_0)$ and $f(x_1)$ we could approximate $f$ on $[x_0, x_1]$ by the polynomial $p_1$ of degree 1 which interpolates $f$ on $\{x_0, x_1\}$. Then an estimate $x_2$ of $x^*$ is obtained by solving the linear equation

$$p_1(x) = 0, \qquad (4.18)$$

whence

$$x_2 = x_1 - \frac{[x_1 - x_0]f(x_1)}{[f(x_1) - f(x_0)]}. \qquad (4.19)$$

If $x^*$ lies between $x_0$ and $x_2$ we approximate $f$ on $[x_0, x_2]$ by the polynomial $p_1$ of degree 1 which interpolates $f$ on $\{x_0, x_2\}$ and again solve Eqn (4.18) to obtain a new estimate $x_3$ of $x^*$. This procedure could be repeated indefinitely thus generating a sequence $\{x_n\}$ which we hope would converge to $x^*$. This procedure is called the *method of false position* and has the advantages that only one evaluation of $f$ per step is needed, and $x^*$ always lies between the next two iterates to be used, thus providing an upper bound for the error at any stage.

The method of false position is given in the following algorithm.

*Algorithm 4.5*

Given $x_0$ and $x_1$ such that $x_0 < x^* < x_1$.

1. Compute $x_2$ from Eqn (4.19) and go to 2.
2. If $f(x_0)f(x_2) < 0$ go to 3; otherwise go to 5.
3. If $|x_0 - x_2| \leqslant \varepsilon$, set $x^* = x_2$ and stop; otherwise go to 4.
4. Set $x_1 = x_2$, and go to 1.
5. If $|x_1 - x_2| \leqslant \varepsilon$, set $x^* = x_2$ and stop; otherwise go to 6.
6. Set $x_0 = x_1$, $x_1 = x_2$ and go to 1. ∎

| $n$ | $x_n$ |
|---|---|
| 0 | 0·7400000 |
| 1 | 0·7384687 |
| 2 | 0·7390834 |
| 3 | 0·7390851 |
| 4 | 0·7390851 |

TABLE 4.5

*Example 4.10*

Table 4.5 shows the results obtained for the equation in Example 4.9 using Algorithm 4.5 with $x_0 = 0.74$, and $x_1 = \cos(0.74)$. This makes it possible to compare the results obtained using Algorithm 4.2 with those obtained using Algorithm 4.5. Using Algorithm 4.5 only *three* iterations are needed to estimate $x^*$ correct to 7D whereas using Algorithm 4.2 *twenty-one* iterations are needed to estimate $x^*$ correct to 6D!

**Tutorial Examples**

1. Show that the following functions satisfy Lipschitz conditions on the intervals given

(1) $\sinh x \quad (\log \tfrac{1}{2} \leqslant x \leqslant \log 2)$;
(2) $1/x^2 \quad (1 \leqslant x \leqslant 2)$;
(3) $|x^3| \quad (-1 \leqslant x \leqslant 1)$;
(4) $\tan^{-1} x \quad (-1 \leqslant x \leqslant 1)$.

2. Apply Lemma 4.3 to the following equations on the intervals given, if possible.

(1) $x = \exp(x) \quad (0 \leqslant x \leqslant 1)$;
(2) $x = 2/x \quad (1 \leqslant x \leqslant 2)$;
(3) $x = \operatorname{cosec} x \quad (0 \leqslant x \leqslant \pi/2)$;
(4) $x = 1 + \tfrac{1}{2} \sin x \quad (0 \leqslant x \leqslant \pi)$.

3. The graph of $f$ is known to cut the $x$-axis at $x^*$ between 0·3 and 0·4 where

$$f(x) = 3x^3 + 5x^2 + x - 1.$$

How many evaluations of $f$ are needed to be sure of estimating the root $x^*$ of the equation

$$f(x) = 0$$

correct to 4D using the method of bisection? Carry out the computation using nested multiplication† to evaluate $f(x)$.

4. By rearrangement if necessary, show that the equation

$$x - \log x - 2 = 0$$

has a unique root $x^*$ in [0·1, 0·2] and give a linearly convergent iterative method for estimating its value. How many iterations are required to ensure an estimate of $x^*$ correct to 6D if $x_0 = 0·1$?

Taking $x_0 = 0·1$, $x_1 = 0·2$ in the method of bisection how many evaluations of $f$ are needed to ensure the same accuracy?

5. It is known that the equation

$$2x = \cos^2 x$$

has a root $x^* = 0·42$ correct to 2D. Use Theorem 4.3 to

† See Tutorial Example 1.2.

construct an iterative method for estimating $x^*$. How many iterations are necessary to ensure 6D accuracy with $x_0 = 0\cdot 42$?

6. Using Theorem 4.4 show that for $x_0$ sufficiently close to $\alpha^{1/2}$, where $\alpha$ is any positive real number, the sequence $\{x_n\}$ generated from
$$x_{n+1} = \frac{(x_n^3 + 3\alpha x_n)}{(3x_n^2 + \alpha)} \quad (n = 0, 1, 2, \ldots)$$
converges to $\alpha^{1/2}$ with order of convergence 3.

7. If $x_0$ is any positive real number and $\{x_n\}$ the Newton sequence generated as in Example 4.8, show that
$$\frac{x_n - \alpha^{1/2}}{x_n + \alpha^{1/2}} = \left[\frac{x_0 - \alpha^{1/2}}{x_0 + \alpha^{1/2}}\right]^{2^n} \quad (n = 0, 1, 2, \ldots),$$
and calculate
$$\lim_{n \to \infty} \frac{(x_{n+1} - \alpha^{1/2})}{(x_n - \alpha^{1/2})^2}.$$
Hence show that $x_n \to \alpha^{1/2}$ as $n \to \infty$ for any positive value of $x_0$.

8. Using the result of Tutorial Example 1.4 and Newton's method, construct an algorithm for estimating a real root of the polynomial equation
$$\sum_{k=0}^{n} a_k x^k = 0,$$
given an initial estimate $x_0$ sufficiently close to the desired root.

Hence compute the root of
$$3x^3 + 5x^2 + x - 1 = 0$$
near $x_0 = 0\cdot 33$, correct to 4D.

9. Apply the secant method with $x_0 = 0\cdot 3$, $x_1 = 0\cdot 33$ and

the method of false position with $x_0 = 0\cdot 3$, $x_1 = 0\cdot 4$ to the equation of Tutorial Example 4.3.

10. Using Theorem 4.3 show that if Eqn (4.1) has a root $x^*$ and the constant $\alpha$ is such that
$$0 < \alpha f^{(1)}(x^*) < 2,$$
with $f^{(1)}$ continuous on an interval $I$ containing $x^*$, then there is a number $\rho$ such that the sequence $\{x_n\}$ generated from
$$x_{n+1} = x_n - \alpha f(x_n) \quad (n = 0, 1, 2, \ldots)$$
converges to $x^*$ if $x^* - \rho \leqslant x_0 \leqslant x^* + \rho$. Use Theorem 4.4 to show that if
$$\alpha = \frac{1}{f^{(1)}(x^*)},$$
then the order of convergence of $\{x_n\}$ is 2.

# CHAPTER 5

# Numerical Solution of Systems of Linear Algebraic Equations

## 5.1 Introduction

The system of linear algebraic equations of Problem 1.5 is expressible in matrix form as

$$\mathbf{A}\mathbf{x} = \mathbf{b}, \tag{5.1}$$

where $\mathbf{A}$ is a square matrix of order $n$, and $\mathbf{x}$ and $\mathbf{b}$ are column vectors, defined by

$$\mathbf{A} = [a_{ij}], \quad \mathbf{x} = [x_j], \quad \mathbf{b} = [b_i] \quad (i, j = 1, \ldots, n). \tag{5.2}$$

Example 1.5 illustrates the notation for $n = 4$.

A *dense* matrix $\mathbf{A}$ has most of its elements $a_{ij}$ non-zero, and a *sparse* matrix $\mathbf{A}$ has most of its elements $a_{ij}$ equal to zero. The elements $a_{ii}$ ($i = 1, \ldots, n$) constitute the *main diagonal* of the square matrix $\mathbf{A}$ of order $n$. If the non-zero elements of $\mathbf{A}$ are located along the main diagonal and on neighbouring diagonals, which are symmetrically distributed about the main diagonal, then $\mathbf{A}$ is called a *band matrix*. Band matrices can be either dense or sparse. In general, the band of non-zero elements contains $(2m + 1)$ diagonals where $0 \leqslant m < n/2$. The matrix $\mathbf{A}$ is called *diagonal* if $m = 0$ and *tri-diagonal* if $m = 1$.

Methods for the numerical solution of Eqn (5.1) are classified as either *direct* methods or *iterative* methods. Direct methods enable Eqn (5.1) to be solved by using a finite number of arithmetical operations, with an accuracy determined by rounding error. Using digital computers, linear systems with dense matrices $\mathbf{A}$ of order up to about 1,000 can be solved using direct methods. For systems of order greater than 1,000 it is essential that **A** be sparse because of number storage problems.

Iterative methods enable a sequence $\{\mathbf{x}^{(i)}\}$ of vectors to be computed which under certain conditions on **A** will converge to the solution vector $\mathbf{x}^*$, as $i \to \infty$, and are relatively unaffected by rounding error. For large-order linear systems such as occur in connection with the numerical solution of partial differential equations, **A** is usually a band matrix and often tridiagonal. For such systems iterative methods are very effective with $n$ as large as 100,000 or more.

## 5.2 Gauss Elimination

All direct methods are based upon the idea of successive elimination of the unknowns $x_k$ ($k = 1, \ldots, n$) due to Gauss, which we illustrate using a system of 3 linear equations.

$$\begin{aligned}
a_{11}x_1 + a_{12}x_2 + a_{13}x_3 &= b_1, \\
a_{21}x_1 + a_{22}x_2 + a_{23}x_3 &= b_2, \\
a_{31}x_1 + a_{32}x_2 + a_{33}x_3 &= b_3.
\end{aligned} \qquad (5.3)$$

Assume that $a_{11} \ne 0$, for if $a_{11} = 0$ we could permute the equations until $a_{11} \ne 0$. If this cannot be done then $a_{i1} = 0$ ($i = 1, 2, 3$) whence the determinant of **A** is zero and so **A** is singular.

Multiply equation 1 in system (5.3) by $-(a_{m1}/a_{11})$ and add the result to equation $m$ ($m = 2, 3$). We obtain the system

$$\begin{aligned}
a_{11}x_1 + a_{12}x_2 + a_{13}x_3 &= b_1, \\
a_{22}^{(1)}x_2 + a_{23}^{(1)}x_3 &= b_2^{(1)}, \\
a_{32}^{(1)}x_2 + a_{33}^{(1)}x_3 &= b_3^{(1)},
\end{aligned} \qquad (5.4)$$

where

$$a_{mj}^{(1)} = a_{mj} - \left(\frac{a_{m1}}{a_{11}}\right)a_{1j}, \quad b_m^{(1)} = b_m - \left(\frac{a_{m1}}{a_{11}}\right)b_1, \quad (5.5)$$

in which $m = 2, 3$ and $j = 2, 3$.

Assume that $a_{22}^{(1)} \ne 0$. Multiply equation 2 in system (5.4) by $-(a_{32}^{(1)}/a_{22}^{(1)})$ and add the result to equation 3. We obtain the system

$$a_{11}x_1 + a_{12}x_2 + a_{13}x_3 = b_1,$$
$$a_{22}^{(1)}x_2 + a_{23}^{(1)}x_3 = b_2^{(1)}, \qquad (5.6)$$
$$a_{33}^{(2)}x_3 = b_3^{(2)},$$

where

$$a_{33}^{(2)} = a_{32}^{(1)} - \left(\frac{a_{32}^{(1)}}{a_{22}^{(1)}}\right)a_{23}^{(1)}, \quad b_3^{(2)} = b_3^{(1)} - \left(\frac{a_{32}^{(1)}}{a_{22}^{(1)}}\right)b_2^{(1)}. \quad (5.7)$$

Now system (5.6) is obtained from system (5.3) by adding to the rows of **A** multiples of other rows of **A**. The determinant of the resulting matrix is equal to the determinant of **A**. Any permutation of the equations permutes the rows of the matrix and changes only the sign of the determinant. Using expansion by rows, it is clear† that the determinant of **A** is $\pm a_{11}a_{22}^{(1)}a_{33}^{(2)}$, the sign depending on whether the equations have been permuted. Hence we have a method for computing the determinant of a non-singular matrix which is far more computationally efficient than a direct expansion.

The system (5.6) is readily solved by *back substitution*. We have

$$x_3 = \frac{b_3^{(2)}}{a_{33}^{(2)}},$$
$$x_2 = \frac{(b_2^{(1)} - a_{23}^{(1)}x_3)}{a_{22}^{(1)}}, \qquad (5.8)$$
$$x_1 = \frac{(b_1 - a_{12}x_2 - a_{13}x_3)}{a_{11}}.$$

Clearly this procedure could in principle be applied to systems of any number of equations.

*Example 5.1*

Consider the system

$$\begin{array}{rl} & 2\cdot 0x_1 + 1\cdot 0x_2 - 2\cdot 0x_3 + 1\cdot 0x_4 = 1\cdot 0, \\ (-1\cdot 0) & 2\cdot 0x_1 + 1\cdot 0x_2 - 4\cdot 0x_3 + 2\cdot 0x_4 = 0\cdot 4, \\ (-1\cdot 5) & 3\cdot 0x_1 - 2\cdot 0x_2 + 3\cdot 0x_3 - 1\cdot 0x_4 = 1\cdot 6, \\ (0\cdot 5) & -1\cdot 0x_1 + 3\cdot 0x_2 - 1\cdot 0x_3 + 1\cdot 0x_4 = 1\cdot 4. \end{array} \qquad (5.9)$$

† See Tutorial Example 5.1.

The numbers in parentheses are the multipliers $-(a_{m1}/a_{11})$ ($m = 2, 3, 4$). Eliminating $x_1$ from the last 3 equations, we obtain after interchanging the second and third of the resulting system of equations

$$\begin{aligned} 2\cdot 0 x_1 + 1\cdot 0 x_2 - 2\cdot 0 x_3 + 1\cdot 0 x_4 &= 1\cdot 0, \\ -3\cdot 5 x_2 + 6\cdot 0 x_3 - 2\cdot 5 x_4 &= 0\cdot 1, \\ (0\cdot 0) \qquad 0\cdot 0 x_2 - 2\cdot 0 x_3 + 1\cdot 0 x_4 &= -0\cdot 6, \\ (1\cdot 0) \qquad 3\cdot 5 x_2 - 2\cdot 0 x_3 + 1\cdot 5 x_4 &= 1\cdot 9. \end{aligned} \qquad (5.10)$$

The numbers in parentheses are the multipliers $-(a^{(1)}_{m2}/a^{(1)}_{22})$ ($m = 3, 4$).

On eliminating $x_2$ from the last 2 equations we obtain

$$\begin{aligned} 2\cdot 0 x_1 + 1\cdot 0 x_2 - 2\cdot 0 x_3 + 1\cdot 0 x_4 &= 1\cdot 0, \\ -3\cdot 5 x_2 + 6\cdot 0 x_3 - 2\cdot 5 x_4 &= 0\cdot 1, \\ -2\cdot 0 x_3 + 1\cdot 0 x_4 &= -0\cdot 6, \\ (2\cdot 0) \qquad 4\cdot 0 x_3 - 1\cdot 0 x_4 &= 2\cdot 0. \end{aligned} \qquad (5.11)$$

The number in parentheses is the multiplier $-(a^{(2)}_{43}/a^{(2)}_{33})$.

Eliminating $x_3$ from the last equation we obtain

$$\begin{aligned} 2\cdot 0 x_1 + 1\cdot 0 x_2 - 2\cdot 0 x_3 + 1\cdot 0 x_4 &= 1\cdot 0, \\ -3\cdot 5 x_2 + 6\cdot 0 x_3 - 2\cdot 5 x_4 &= 0\cdot 1, \\ -2\cdot 0 x_3 + 1\cdot 0 x_4 &= -0\cdot 6, \\ 1\cdot 0 x_4 &= 0\cdot 8. \end{aligned}$$

Back substitution then gives $x_4 = 0\cdot 8$, $x_3 = 0\cdot 7$, $x_2 = 0\cdot 6$, $x_1 = 0\cdot 5$. The determinant of $\mathbf{A}$ is given by

$$\text{Det}(\mathbf{A}) = -a_{11} a^{(1)}_{22} a^{(2)}_{33} a^{(3)}_{44},$$

where the minus sign is due to the interchange of equations resulting in the system (5.10). Since $a_{11} = 2\cdot 0$, $a^{(1)}_{22} = -3\cdot 5$, $a^{(2)}_{33} = -2\cdot 0$, and $a^{(3)}_{44} = 1\cdot 0$, we have $\text{Det}(\mathbf{A}) = -14$. The reader should verify by direct expansion that this value of $\text{Det}(\mathbf{A})$ is correct.

## 5.3 Choice of Pivots

In Section 5.2, $a_{11}$, $a^{(1)}_{22}$, $a^{(2)}_{33}$, $a^{(3)}_{44}$ are called *pivots*. It is extremely important to avoid the use of pivots $a^{(k-1)}_{kk}$ for which $|a_{kk}^{(k-1)}|$ is small compared with $|a_{ik}^{(k-1)}|$ $(i = k+1, \ldots, n)$. To understand why this is so, consider the system

$$ax + by = p,$$
$$cx + dy = q, \qquad (5.12)$$

for which the pivot is $a$. Solving the system we obtain

$$y = \frac{\left(q - \frac{c}{a}p\right)}{\left(d - \frac{c}{a}b\right)}, \quad x = \frac{p}{a} - \frac{b}{a}y. \qquad (5.13)$$

Suppose that the coefficients $a$, $b$, $c$, $d$, $p$, and $q$ are varied by $\delta a$, $\delta b$, $\delta c$, $\delta d$, $\delta p$, and $\delta q$ respectively. Then by Tutorial Example 1.8 it is clear that if $|a|$ is small compared with $|b|$, $|c|$, $|d|$, $|p|$, and $|q|$, then the corresponding variations $\delta x$ and $\delta y$ in $x$ and $y$ could be very large. If, however, $|a|$ is not less that $|b|$, $|c|$, $|d|$, $|p|$, and $|q|$ then $|\delta x|$ and $|\delta y|$ are not unduly great. Even if the exact values of $a$, $b$, $c$, $d$, $p$, and $q$ are used, it is still important to avoid the use of $a$ as a pivot if $|a|$ is small compared with the magnitudes of the other coefficients, especially when using a digital computer, in which floating point arithmetic of fixed precision is often used. This means that all numbers are expressed in the form $\pm 0 \cdot n_1 \ldots n_k \times 10^{\pm m}$ where $n_1, \ldots, n_k$ are all positive integers each of which can take the values $0, \ldots, 9$ and $m$ is a positive integer. The number $\pm 0 \cdot n_1 \ldots, n_k$ is called the *mantissa* and $\pm m$ is the *exponent*. The number $k$ determines the precision. If two floating point numbers are multiplied, the mantissa of the product contains not less than $2k - 1$ digits, which must be rounded to $k$ digits, thereby introducing a rounding error of magnitude at most $0 \cdot 5 \times 10^{\pm m-k}$. The rounding errors produced in forming $(c/a)p$, $(c/a)b$ and $(b/a)y$ in Eqn (5.13) are magnified if $|a|$ is small compared with $|b|$, $|c|$, and $|p|$. A detailed analysis of Gauss elimination shows that in general it is advisable to keep the numbers $|a_{ik}^{(k-1)}/a_{kk}^{(k-1)}|$ $(i, k = 1, \ldots, n)$ not greater than unity. One method of ensuring this is the so-called *partial pivoting* strategy of Wilkinson, in which we interchange equa-

## §5.3] CHOICE OF PIVOTS

tions $i$ and $l$ after determining that $a_{il}^{(l-1)}$ is the element of greatest magnitude in column $l$.

*Example 5.2*

Consider the system

$$(0\cdot100)10^{-3}x_1 + (0\cdot100)10^1 x_2 = (0\cdot200)10^1,$$
$$(0\cdot100)10^1 x_1 + (0\cdot100)10^1 x_2 = (0\cdot300)10^1,$$

in which we have used a floating point representation of the coefficients with $k = 3$. Using exact arithmetic we find that

$$x_1 = \frac{10{,}000}{9{,}999}, \quad x_2 = \frac{19{,}997}{9{,}999},$$

so $x_1 = 1\cdot0001$, $x_2 = 2\cdot0000$ correct to 4D.

Let us solve the system using finite precision floating point arithmetic as would be done in a digital computer. To understand the computation we must realize that, for example,

$$(0\cdot100)10^5 - (0\cdot100)10^1 = 10{,}000 - 1 = 9{,}999 = (0\cdot100)10^5$$

to the degree of precision used.

Using $(0\cdot100)10^{-3}$ as a pivot, we eliminate $x_1$ from equation 2 and obtain

$$(0\cdot100)10^{-3}x_1 + (0\cdot100)10^1 x_2 = (0\cdot200)10^1,$$
$$(-0\cdot100)10^5 x_2 = (-0\cdot200)10^5.$$

Hence by back substitution,

$$x_2 = \frac{(-0\cdot200)}{(-0\cdot100)} = (0\cdot200)10^1,$$
$$x_1 = \frac{[(0\cdot200)10^1 - (0\cdot200)10^1]}{(0\cdot100)10^{-3}} = (0\cdot000)10^{-2}.$$

So we obtain $x_1 = 0\cdot00$, $x_2 = 2\cdot00$ correct to 2D. Although the estimate of $x_2$ is satisfactory, that of $x_1$ is completely wrong!

We now solve the system using fixed precision floating point arithmetic and Wilkinson's partial pivoting strategy. Considering the original system of equations we note that

$|a_{21}| = (0\cdot100)10^1$ is greater than $|a_{11}| = (0\cdot100)10^{-3}$

so we interchange equations 1 and 2 so that the new pivot is $(0\cdot100)10^1$, the element of greatest magnitude in column 1. Eliminating $x_1$ from the second equation of the permuted system we obtain

$$(0\cdot100)10^1 x_1 + (0\cdot100)10^1 x_2 = (0\cdot300)10^1,$$
$$(0\cdot100)10^1 x_2 = (0\cdot200)10^1,$$

whence

$$x_2 = \frac{(0\cdot200)10^1}{(0\cdot100)10^1} = (0\cdot200)10^1,$$

$$x_1 = \frac{[(0\cdot300)10^1 - (0\cdot200)10^1]}{(0\cdot100)10^1} = (0\cdot100)10^1.$$

So $x_1 = 1\cdot00$, $x_2 = 2\cdot00$ correct to 2D, which is completely satisfactory.

Gauss elimination with partial pivoting, or one of its many variants† is probably the most effective method for the numerical solution of linear systems with dense matrices of small order.

## 5.4 Computation of Inverse Matrices

Gauss elimination with partial pivoting can be used to estimate the inverse of a non-singular square matrix $\mathbf{A}$ of order $n$. Let $\mathbf{X}$ be a square matrix of order $n$ with elements $x_{ij}$, and let

$$\mathbf{AX} = \mathbf{I}, \qquad (5.14)$$

where $\mathbf{I}$ is the unit matrix of order $n$. Then Eqn (5.14) may be written

$$\sum_{k=1}^{n} a_{ik} x_{kj} = \delta_{ij} \quad (i, j = 1, \ldots, n). \qquad (5.15)$$

Let

$$\mathbf{x}_j = [x_{1j}, \ldots, x_{nj}]^T, \quad \mathbf{e}_j = [0, \ldots, 0, \overset{(j)}{1}, 0, \ldots, 0]^T, \qquad (5.16)$$

so that $\mathbf{x}_j$ is a column vector the elements of which constitute

† See, for example, Noble (1964), National Physical Laboratory (1962), Beckett and Hurt (1967), Ralston (1965), Ralston and Wilf (1966).

column $j$ of $\mathbf{X}$, and $\mathbf{e}_j$ is a column vector with elements the values of which are all zero save that of the $j$th, which is unity. Then Eqns (5.15) are equivalent to the set of $n$ linear systems

$$\mathbf{A}\mathbf{x}_j = \mathbf{e}_j \quad (j = 1, \ldots, n). \tag{5.17}$$

We can solve these $n$ linear systems for the $n$ columns of $\mathbf{X}$ and hence obtain $\mathbf{X}$. But from Eqn (5.14) it is clear that $\mathbf{X} = \mathbf{A}^{-1}$. Hence we have a method of estimating the elements of $\mathbf{A}^{-1}$ by solving the $n$ linear systems (5.17). Note that very little more computational labour is required for solving these $n$ systems than is required for solving just one such system, for all the systems have the same matrix $\mathbf{A}$.

*Example 5.3*

Let

$$\mathbf{A} = \begin{bmatrix} 0\cdot866, & -0\cdot500, & 0\cdot000 \\ 0\cdot500, & 0\cdot866, & 0\cdot000 \\ 0\cdot000, & 0\cdot000, & 1\cdot000 \end{bmatrix}.$$

Then

$$\mathbf{e}_1 = \begin{bmatrix} 1 \\ 0 \\ 0 \end{bmatrix}, \quad \mathbf{e}_2 = \begin{bmatrix} 0 \\ 1 \\ 0 \end{bmatrix}, \quad \mathbf{e}_3 = \begin{bmatrix} 0 \\ 0 \\ 1 \end{bmatrix},$$

and we must solve the three systems

$$\mathbf{A}\mathbf{x}_j = \mathbf{e}_j \quad (j = 1, 2, 3).$$

Since each system has the same matrix $\mathbf{A}$ we solve all three simultaneously, writing three right-hand sides corresponding to $j = 1, 2, 3$.

We have

$$\begin{aligned} 0\cdot866x_{1j} - 0\cdot500x_{2j} + 0\cdot000x_{3j} &= 1 \quad 0 \quad 0, \\ 0\cdot500x_{1j} + 0\cdot866x_{2j} + 0\cdot000x_{3j} &= 0 \quad 1 \quad 0, \\ 0\cdot000x_{1j} + 0\cdot000x_{2j} + 1\cdot000x_{3j} &= 0 \quad 0 \quad 1. \end{aligned}$$

We need no interchanges, and only one elimination is required, whence we obtain for the reduced system

$$\begin{aligned} 0\cdot866x_{1j} - 0\cdot500x_{2j} + 0\cdot000x_{3j} &= \phantom{-}1 \phantom{.57737} \quad 0 \quad 0, \\ 1\cdot15468x_{2j} + 0\cdot000x_{3j} &= -0\cdot57737 \quad 1 \quad 0, \\ 1\cdot000x_{3j} &= \phantom{-}0 \phantom{.57737} \quad 0 \quad 1. \end{aligned}$$

So by back substitution we obtain

$$x_{31} = 0.000, \quad x_{32} = 0.000, \quad x_{33} = 1.000,$$
$$x_{21} = -0.500, \quad x_{22} = 0.866, \quad x_{23} = 0.000,$$
$$x_{11} = 0.866, \quad x_{12} = 0.500, \quad x_{13} = 0.000,$$

whence

$$\mathbf{A}^{-1} = \begin{bmatrix} 0.866 & 0.500 & 0.000 \\ -0.500 & 0.866 & 0.000 \\ 0.000 & 0.000 & 1.000 \end{bmatrix},$$

correct to 3D. It is readily verified that this is correct, for $\mathbf{A}$ is an orthogonal matrix the inverse of which is its transpose.

In general the inversion of square matrices of large order should be avoided if possible because of rounding errors. Where it is necessary to estimate $\mathbf{A}^{-1}$, the method described involves far less computational labour than a direct estimate using the analytical expression for $\mathbf{A}^{-1}$.

## 5.5 Conditioning

Consider the linear systems

$$\begin{aligned} 100x_1 + 99x_2 &= 398, \\ 99x_1 + 98x_2 &= 394, \end{aligned} \quad (5.18)$$

the exact solution of which is $x_1 = 2$, $x_2 = 2$, and

$$\begin{aligned} 100x_1 + 99x_2 &= 398, \\ 99x_1 + 98x_2 &= 393.98, \end{aligned}$$

the exact solution of which is $x_1 = -0.02$, $x_2 = 4$.

If we regard the second system as having been obtained from the first by a small variation of the right-hand side, then we see that the numerical solution of the first system is extremely sensitive to such small variations.

Consider the two linear systems

$$\begin{aligned} x_1 + x_2 &= 4, \\ 1.01x_1 + x_2 &= 3.02, \end{aligned} \quad (5.19)$$

which has the exact solution $x_1 = 2$, $x_2 = 2$, and

$$\begin{aligned} x_1 + x_2 &= 4, \\ 1.06125x_1 + x_2 &= 3.02, \end{aligned}$$

which has the exact solution $x_1 = -16$, $x_2 = 20$. We see that a small variation in the matrix of the first system produces a very large variation in the numerical solution of the system. Systems (5.18) and (5.19) are examples of *ill-conditioning*. The system (5.1) is said to be *ill-conditioned* when small relative variations in the elements of **A** or **b** produce large relative variations in the solution vector **x**.

If **A** or **b** is not known exactly as is the case when, for example, they are obtained from a physical experiment by measurement, then the solution vector **x** of Eqn (5.1) can be very much in error, no matter how accurately the computation of **x** is performed. Also, the values of the elements **A** and **b** must often be rounded for storage in a computing machine, again producing uncertainties in **A** and **b** which for an ill-conditioned system can lead to serious error in the solution.

## 5.6 Iterative Improvement of Solutions

If **A** and **b** are obtained empirically and the system is ill-conditioned, then nothing can be done to improve the accuracy of a solution of Eqn (5.1). If however **A** and **b** are known exactly but must be rounded for use in a computing machine then the resulting error can be reduced even for an ill-conditioned system by using a larger number of significant figures both for the elements of **A** and **b** and for the computation which yields the solution. The use of multiple precision in a digital computer (that is, using two or three times the usual number of significant figures) greatly increases the time required for the computation and is therefore undesirable. Where rounding error is the cause of inaccuracy, a solution $\mathbf{x}^{(0)}$ computed using single precision arithmetic can often be improved by using the *residual* vector $\mathbf{r}^{(0)}$ defined by

$$\mathbf{r}^{(0)} = \mathbf{b} - \mathbf{A}\mathbf{x}^{(0)}. \quad (5.20)$$

For systems which are not ill-conditioned the smallness of $\mathbf{r}^{(0)}$ is a good indication of the accuracy of the solution $\mathbf{x}^{(0)}$, but this is not so for ill-conditioned systems. However, if $\mathbf{y}^{(0)}$ is the exact solution of

$$\mathbf{A}\mathbf{y}^{(0)} = \mathbf{r}^{(0)}, \tag{5.21}$$

and

$$\mathbf{x}^{(1)} = \mathbf{x}^{(0)} + \mathbf{y}^{(0)}, \tag{5.22}$$

then

$$\mathbf{A}\mathbf{x}^{(1)} = \mathbf{A}\mathbf{x}^{(0)} + \mathbf{A}\mathbf{y}^{(0)} = \mathbf{A}\mathbf{x}^{(0)} + \mathbf{r}^{(0)} = \mathbf{b},$$

so $\mathbf{x}^{(1)}$ is the exact solution of the system to be solved. Of course, Eqn (5.21) cannot be solved exactly because of rounding errors, but if $\mathbf{r}^{(0)}$ were computed with sufficient precision we would expect $\mathbf{x}^{(1)}$ given by Eqn (5.22) to be a better estimate of the solution of Eqn (5.1) than is $\mathbf{x}^{(0)}$. It is *vital* that $\mathbf{r}^{(0)}$ be computed with greater precision than that of the rest of the computation. The fact that $\mathbf{y}^{(0)}$ need not be computed more accurately than $\mathbf{x}^{(0)}$ means that very little additional labour is needed to compute $\mathbf{y}^{(0)}$ once $\mathbf{x}^{(0)}$ has been computed because equations (5.1) and (5.21) have the same matrix $\mathbf{A}$. If the given system is not too badly conditioned, improvement can be continued iteratively by repeating the procedure with $\mathbf{x}^{(0)}$ replaced by $\mathbf{x}^{(1)}$ etc., whence we obtain the following algorithm.

*Algorithm 5.1*

1. Using double precision, compute $\mathbf{r}^{(0)}$ from Eqn (5.20) and go to 2.
2. Using single precision, compute $\mathbf{y}^{(0)}$ from Eqn (5.21) and go to 3.
3. Using single precision, compute $\mathbf{x}^{(1)}$ from Eqn (5.22) and go to 4.
4. If $\mathbf{x}^{(1)}$ satisfies a given convergence criterion set $\mathbf{x} = \mathbf{x}^{(1)}$ and stop; otherwise go to 5.
5. Set $\mathbf{x}^{(0)} = \mathbf{x}^{(1)}$ and go to 1.∎

*Example 5.4*

Using floating point arithmetic with $k = 4$, Gauss elimination when applied to Eqn (5.19) gives

$$x_1^{(0)} = (0 \cdot 3980)10^1, \quad x_2^{(0)} = (0 \cdot 0000)10^0.$$

Using double precision arithmetic ($k = 8$),

$$\mathbf{A}\mathbf{x}^{(0)} = \begin{bmatrix} (0\cdot 39800000)10^3 \\ (0\cdot 39402000)10^3 \end{bmatrix}, \quad \mathbf{r}^{(0)} = \begin{bmatrix} (0\cdot 00000000)10^0 \\ (-0\cdot 00002000)10^3 \end{bmatrix}.$$

Using single precision arithmetic, $y_1^{(0)} = (-0\cdot 1980)10^1$, $y_2^{(0)} = (0\cdot 2000)10^1$, so $x_1^{(1)} = (0\cdot 2000)10^1$, $x_2^{(1)} = (0\cdot 2000)10^1$, which agrees with the exact solution to 3D.

For a much more comprehensive treatment of iterative improvement and, indeed, of the whole subject of the numerical solution of systems of linear algebraic equations using a computer the more advanced reader should consult Forsythe and Moler (1967).

## 5.7 The Jacobi Method

Equation (5.1) can be expressed in the form

$$x_k^* = \frac{1}{a_{kk}} \left[ b_k - \sum_{\substack{j=1 \\ j \neq k}}^{n} a_{kj} x_j^* \right] \quad (k = 1, \ldots, n), \qquad (5.23)$$

if $a_{kk} \neq 0$ ($k = 1, \ldots, n$), where $\mathbf{x}^*$ is the solution vector. Given an estimate $\mathbf{x}^{(0)}$ of $\mathbf{x}^*$ we can generate a sequence $\{\mathbf{x}^{(i)}\}$ of vectors from

$$x_k^{(i+1)} = \frac{1}{a_{kk}} \left[ b_k - \sum_{\substack{j=1 \\ j \neq k}}^{n} a_{kj} x_j^{(i)} \right]$$
$$(k = 1, \ldots, n)\ (i = 0, 1, 2, \ldots). \qquad (5.24)$$

Write

$$||\mathbf{x}^* - \mathbf{x}^{(i)}|| = \max_{1 \leq k \leq n} \{|x_k^* - x_k^{(i)}|\}$$
$$(i = 0, 1, 2 \ldots), \qquad (5.25)$$

and

$$\alpha = \max_{1 \leq k \leq n} \left\{ \sum_{\substack{j=1 \\ j \neq k}}^{n} \left|\frac{a_{kj}}{a_{kk}}\right| \right\}. \qquad (5.26)$$

Then subtracting Eqn (5.24) from Eqn (5.23), taking the

modulus of the result, and using the triangle inequality, we obtain

$$|x_k^* - x_k^{(i+1)}| \leqslant \sum_{\substack{j=1 \\ \neq k}}^{n} \left|\frac{a_{kj}}{a_{kk}}\right| |x_j^* - x_j^{(i)}| \quad (k = 1, \ldots, n)$$

$$\leqslant \|\mathbf{x}^* - \mathbf{x}^{(i)}\| \sum_{\substack{j=1 \\ j \neq k}}^{n} \left|\frac{a_{kj}}{a_{kk}}\right| \quad (k = 1, \ldots, n).$$

So

$$\|\mathbf{x}^* - \mathbf{x}^{(i+1)}\| \leqslant \alpha \|\mathbf{x}^* - \mathbf{x}^{(i)}\|, \quad (i = 0, 1, 2, \ldots).$$

By repeated use of this inequality we obtain

$$\|\mathbf{x}^* - \mathbf{x}^{(i+1)}\| \leqslant \alpha^{i+1} \|\mathbf{x}^* - \mathbf{x}^{(0)}\|, \quad (i = 0, 1, 2, \ldots).$$

If $0 \leqslant \alpha < 1$ we therefore have $\|\mathbf{x}^* - \mathbf{x}^{(i+1)}\| \to 0$ as $i \to \infty$.

Hence from Eqn (5.25),

$$|x_k^* - x_k^{(i+1)}| \to 0 \quad (k = 1, \ldots, n) \quad \text{as } i \to \infty.$$

We therefore see that, starting with *any* initial estimate $\mathbf{x}^{(0)}$ of the solution vector $\mathbf{x}^*$ of Eqn (5.1) the sequence of vectors $\{\mathbf{x}^{(i)}\}$ generated from Eqn (5.24) converges to $\mathbf{x}^*$ if only $\alpha < 1$. From Eqn (5.26), if $\alpha < 1$ then

$$\sum_{\substack{j=1 \\ j \neq k}}^{n} |a_{kj}| < |a_{kk}| \quad (k = 1, \ldots, n). \tag{5.27}$$

A matrix $\mathbf{A}$ the elements of which satisfy the inequalities (5.27) is called *strictly diagonally dominant*. We have therefore proved the following theorem.

### THEOREM 5.1

If $\mathbf{A}$ is strictly diagonally dominant, then the sequence of vectors $\{\mathbf{x}^{(i)}\}$ generated from Eqn (5.24) with $\mathbf{x}^{(0)}$ chosen arbitrarily, converges to the solution $\mathbf{x}^*$ of Eqn (5.1). ∎

*Example 5.5*

Consider the system

$$\begin{aligned} 10x_1 + x_2 + x_3 &= 15, \\ x_1 + 10x_2 + x_3 &= 24, \\ x_1 + x_2 + 10x_3 &= 33. \end{aligned}$$

For this system
$$|a_{12}| + |a_{13}| = 2 < |a_{11}| = 10,$$
$$|a_{21}| + |a_{23}| = 2 < |a_{22}| = 10,$$
$$|a_{31}| + |a_{32}| = 2 < |a_{33}| = 10,$$

so the matrix **A** of the system is strictly diagonally dominant. If no initial estimate is available it is advisable to use $\mathbf{x}^{(0)} = \mathbf{0}$, unless $\alpha \ll 1$, when it is advisable to take

$$x_k^{(0)} = \frac{b_k}{a_{kk}} \quad (k = 1, \ldots, n). \tag{5.28}$$

We therefore take

$$x_1^{(0)} = 1\cdot 5, \quad x_2^{(0)} = 2\cdot 4, \quad x_3^{(0)} = 3\cdot 3,$$

and generate the sequence $\{\mathbf{x}^{(i)}\}$ from

$$x_1^{(i+1)} = \frac{1}{a_{11}}[b_1 - a_{12}x_2^{(i)} - a_{13}x_3^{(i)}] = \frac{1}{10}[15 - x_2^{(i)} - x_3^{(i)}],$$

$$x_2^{(i+1)} = \frac{1}{a_{12}}[b_2 - a_{21}x_1^{(i)} - a_{23}x_3^{(i)}] = \frac{1}{10}[24 - x_1^{(i)} - x_3^{(i)}],$$

$$x_3^{(i+1)} = \frac{1}{a_{33}}[b_3 - a_{31}x_1^{(i)} - a_{32}x_2^{(i)}] = \frac{1}{10}[33 - x_1^{(i)} - x_2^{(i)}].$$

The results obtained are shown in Table 5.1. The exact solution of the system is $x_1 = 1$, $x_2 = 2$, $x_3 = 3$.

| $i$ | $x_1^{(i)}$ | $x_2^{(i)}$ | $x_3^{(i)}$ |
|---|---|---|---|
| 0 | 1·500000 | 2·400000 | 3·300000 |
| 1 | 0·930000 | 1·920000 | 2·910000 |
| 2 | 1·017000 | 2·016000 | 3·015000 |
| 3 | 0·996900 | 1·996800 | 2·996700 |
| 4 | 1·000650 | 2·000640 | 3·000630 |
| 5 | 0·999873 | 1·999872 | 2·999871 |
| 6 | 1·000026 | 2·000026 | 3·000026 |
| 7 | 0·999995 | 1·999995 | 2·999995 |

TABLE 5.1

The Jacobi method for the numerical solution of Eqn (5.1) may be expressed algorithmically as follows, $\varepsilon$ being a given small parameter.

*Algorithm 5.2*

1. Compute $x_k^{(1)}$ ($k = 1, \ldots, n$) from Eqn (5.24) with $i = 0$, and go to 2.
2. If
$$|x_k^{(1)} - x_k^{(0)}| \leqslant \varepsilon \quad (k = 1, \ldots, n),$$
set $x_k^* = x_k^{(1)}$ ($k = 1, \ldots, n$) and stop; otherwise go to 3.
3. Set $x_k^{(0)} = x_k^{(1)}$ ($k = 1, \ldots, n$) and go to 1. ∎

## 5.8 The Gauss–Seidel Method

On examining Eqn (5.24) for some $k > 1$ we note that at the stage when $x_k^{(i+1)}$ is to be computed, the values of $x_j^{(i+1)}$ ($j = 1, \ldots, k-1$) have already been obtained. It seems natural, therefore, to use the value of $x_j^{(i+1)}$ rather than that of $x_j^{(i)}$ ($j = 1, \ldots, k-1$) in computing $x_k^{(i+1)}$, and we should then intuitively expect more rapid convergence than in the Jacobi method. In place of Eqn (5.24) we then obtain

$$x_1^{(i+1)} = \frac{1}{a_{11}}\left[b_1 - \sum_{j=2}^{n} a_{1j} x_j^{(i)}\right],$$

$$x_k^{(i+1)} = \frac{1}{a_{kk}}\left[b_k - \sum_{j=1}^{k-1} a_{kj} x_j^{(i+1)} - \sum_{j=k+1}^{n} a_{kj} x_j^{(i)}\right],$$
$$(k = 2, \ldots, n-1), \quad (5.29)$$

$$x_n^{(i+1)} = \frac{1}{a_{nn}}\left[b_n - \sum_{j=1}^{n-1} a_{kj} x_j^{(i+1)}\right].$$

To show that the sequence $\{\mathbf{x}^{(i)}\}$ thus generated converges to the solution $\mathbf{x}^*$ of Eqn (5.1), we subtract Eqns (5.29) from the same equations with all $x_s^{(i)}$ and $x_s^{(i+1)}$ replaced by $x_s^*$ ($s = 1, \ldots, n$) as in the proof of Theorem 5.1, and use the triangle inequality. We obtain

$$|x_1^* - x_1^{(i+1)}| \leqslant \sum_{j=2}^{n} \left|\frac{a_{1j}}{a_{11}}\right| |x_j^* - x_j^{(i)}| \leqslant \alpha \|\mathbf{x}^* - \mathbf{x}^{(i)}\|$$
$$(i = 0, 1, 2, \ldots), \quad (5.30)$$

## §5.8]    THE GAUSS–SEIDEL METHOD

$$|x_k^* - x_k^{(i+1)}| \leqslant \sum_{j=1}^{k-1} \left|\frac{a_{kj}}{a_{kk}}\right| |x_j^* - x_j^{(i+1)}|$$
$$+ \sum_{j=k+1}^{n} \left|\frac{a_{kj}}{a_{kk}}\right| |x_j^* - x_j^{(i)}|, \quad (5.31)$$

where $k = 2, \ldots, n - 1$ and $i = 0, 1, 2, \ldots$, and

$$|x_n^* - x_n^{(i+1)}| \leqslant \sum_{j=1}^{n-1} \left|\frac{a_{kj}}{a_{kk}}\right| |x_j^* - x_j^{(i+1)}|,$$
$$(i = 0, 1, 2, \ldots). \quad (5.32)$$

Using Eqn (5.30) in Eqn (5.31) with $k = 2$ we easily obtain

$$|x_2^* - x_2^{(i+1)}| \leqslant \alpha ||\mathbf{x}^* - \mathbf{x}^{(i)}|| \quad (i = 0, 1, 2, \ldots). \quad (5.33)$$

If we then assume that

$$|x_k^* - x_k^{(i+1)}| \leqslant \alpha ||\mathbf{x}^* - \mathbf{x}^{(i)}||, \quad (i = 0, 1, 2, \ldots),$$

for $k = 1, \ldots, m$ with $m < n - 1$ it is easy to prove by induction, using Eqn (5.31), that inequality (5.33) is true for $k = 1, \ldots, n - 1$. Also from Eqn (5.32), the inequality is also true for $k = n$. Hence from Eqn (5.25)

$$||\mathbf{x}^* - \mathbf{x}^{(i+1)}|| \leqslant \alpha ||\mathbf{x}^* - \mathbf{x}^{(i)}|| \quad (i = 0, 1, 2, \ldots). \quad (5.34)$$

Proceeding as for the Jacobi method we then establish the following theorem.

### THEOREM 5.2

If $A$ is strictly diagonally dominant then the sequence of vectors $\{\mathbf{x}^{(i)}\}$ generated from Eqns (5.29) with $\mathbf{x}^{(0)}$ chosen arbitrarily, converges to the solution $\mathbf{x}^*$ of Eqn (5.1). ∎

### Example 5.6

For the linear system given in Example 5.5, Eqns (5.29) become

$$x_1^{(i+1)} = \frac{1}{a_{11}}[b_1 - a_{12}x_2^{(i)} - a_{13}x_3^{(i)}]$$
$$= \frac{1}{10}[15 - x_2^{(i)} - x_3^{(i)}],$$

$$x_2^{(i+1)} = \frac{1}{a_{22}}[b_2 - a_{21}x_1^{(i+1)} - a_{23}x_3^{(i)}]$$

$$= \frac{1}{10}[24 - x_1^{(i+1)} - x_3^{(i)}],$$

$$x_3^{(i+1)} = \frac{1}{a_{33}}[b_3 - a_{31}x_1^{(i+1)} - a_{32}x_2^{(i+1)}]$$

$$= \frac{1}{10}[33 - x_1^{(i+1)} - x_2^{(i+1)}].$$

Again we can use Eqns (5.28) to obtain the initial iterate. The results are shown in Table 5.2.

| $i$ | $x_1^{(i)}$ | $x_2^{(i)}$ | $x_3^{(i)}$ |
|---|---|---|---|
| 0 | 1·500000 | 2·400000 | 3·300000 |
| 1 | 0·930000 | 1·977000 | 3·009300 |
| 2 | 1·001370 | 1·998933 | 2·999970 |
| 3 | 1·000110 | 1·999992 | 2·999990 |
| 4 | 1·000002 | 2·000001 | 3·000000 |

TABLE 5.2

The Gauss–Seidel method may be expressed algorithmically as follows.

*Algorithm 5.3*

1. Compute $x_k^{(1)}$ ($k = 1, \ldots, n$) from Eqns (5.29) with $i = 0$, and go to 2.
2. If
$$|x_k^{(1)} - x_k^{(0)}| \leqslant \varepsilon \quad (k = 1, \ldots, n),$$
set $x_k^* = x_k^{(1)}$ ($k = 1, \ldots, n$) and stop; otherwise go to 3.
3. Set $x_k^{(0)} = x_k^{(1)}$ ($k = 1, \ldots, n$) and go to 1.

It is not in general true that the convergence of the Jacobi method implies that of the Gauss–Seidel method, nor is the converse generally true, but when **A** satisfies Eqn (5.27), the

## Tutorial Examples

1. Using expansion by rows, show that if $\mathbf{A}$ is a lower-triangular matrix given by

$$\mathbf{A} = [a_{ij}]; \quad a_{ij} = 0 \quad (i < j) \, (i, j = 1, \ldots, n),$$

then

$$\text{Det}(\mathbf{A}) = a_{11} a_{22} \ldots a_{nn}.$$

By noting that

$$\text{Det}(\mathbf{A}^T) = \text{Det}(\mathbf{A})$$

for any particular square matrix $\mathbf{A}$, show that if $\mathbf{A}$ is an upper-triangular matrix given by

$$\mathbf{A} = [a_{ij}]; \quad a_{ij} = 0 \quad (i > j) \, (i, j = 1, \ldots, n),$$

then again

$$\text{Det}(\mathbf{A}) = a_{11} a_{22} \ldots a_{nn}.$$

2. If

$$\mathbf{A} = \begin{bmatrix} 1{\cdot}00 & 2{\cdot}25 & -3{\cdot}46 & 2{\cdot}67 \\ 2{\cdot}63 & -3{\cdot}40 & 1{\cdot}52 & -3{\cdot}23 \\ 10{\cdot}34 & 2{\cdot}41 & -4{\cdot}62 & 6{\cdot}73 \\ 5{\cdot}63 & -3{\cdot}86 & 3{\cdot}87 & -2{\cdot}44 \end{bmatrix}, \quad \mathbf{b} = \begin{bmatrix} 0{\cdot}5906 \\ 0{\cdot}2423 \\ 8{\cdot}9344 \\ 4{\cdot}2441 \end{bmatrix}$$

use Gauss elimination with partial pivoting and floating point arithmetic with 4 significant figures to estimate the numerical solution of the corresponding linear system. Also use Gauss elimination to estimate $\text{Det}(\mathbf{A})$ and $\mathbf{A}^{-1}$. Compute the residual vector and use it to improve the numerical solution obtained by Gauss elimination.

3. Repeat Tutorial Example 5.2 with $\mathbf{A}$ and $\mathbf{b}$ replaced by

$$\mathbf{A} = \begin{bmatrix} 1 & \frac{1}{2} & \frac{1}{3} \\ \frac{1}{2} & \frac{1}{3} & \frac{1}{4} \\ \frac{1}{3} & \frac{1}{4} & \frac{1}{5} \end{bmatrix}, \quad \mathbf{b} = \begin{bmatrix} 1 \\ 1 \\ 1 \end{bmatrix}.$$

4. Using the triangle inequality for numbers, show that, if **x** and **y** are any two column vectors with $n$ elements, then

$$||\mathbf{x} + \mathbf{y}|| \leq ||\mathbf{x}|| + ||\mathbf{y}||,$$

where

$$||\mathbf{x}|| = \max_{1 \leq k \leq n} \{|x_k|\}.$$

Hence using a similar argument to that used to establish Theorem 4.2, show that for both the Jacobi and Gauss–Seidel methods, with the notation of Sections 5.8 and 5.9,

$$||\mathbf{x}^* - \mathbf{x}^{(i)}|| \leq \frac{\alpha^i}{(1-\alpha)}||\mathbf{x}^{(1)} - \mathbf{x}^{(0)}|| \quad (i = 0, 1, 2, \ldots).$$

5. Rearrange the system of equations

$$\begin{aligned}
2x_1 + 12x_2 + x_3 - x_4 &= 0\cdot 30, \\
4x_1 - 2x_2 + x_3 + 20x_4 &= 62\cdot 80, \\
x_1 + 3x_2 - 24x_3 + 2x_4 &= -53\cdot 75, \\
16x_1 - x_2 + x_3 + 2x_4 &= 10\cdot 05,
\end{aligned}$$

in order that the Jacobi and Gauss–Seidel methods can be used.

With

$$x_1^{(0)} = \frac{10\cdot 05}{16}, \quad x_2^{(0)} = \frac{0\cdot 30}{12}, \quad x_3^{(0)} = \frac{53\cdot 75}{24}, \quad x_4^{(0)} = \frac{62\cdot 80}{20}$$

how many Jacobi iterations are required to ensure a solution accurate to 6D?

CHAPTER 6

# The Numerical Solution of First-Order Ordinary Differential Equations

## 6.1 Introduction

Many of the laws governing change in Nature are expressible in terms of differential equations which have no known analytical solutions suitable for numerical calculation. In this chapter we consider the numerical solution of first-order ordinary differential equations with initial conditions, which are of the form

$$y^{(1)}(x) = f(x, y(x)), \quad y(a) = A, \quad (a \leqslant x \leqslant b), \qquad (6.1)$$

in which $f$ is defined for $a \leqslant x \leqslant b$, $c \leqslant y \leqslant d$, and the initial value $A$ of $y$ is given. Sufficient conditions for the existence of a unique solution function $y$ of Eqn (6.1) are given in the following theorem.

*THEOREM 6.1*

Let $D$ be a rectangular domain

$$a \leqslant x \leqslant b, \quad A - c \leqslant y \leqslant A + c.$$

If

(1) $f$ is continuous in $D$ with $|f(x, y)| \leqslant M$ for all points $(x, y)$ in $D$, $M$ being a given number;

(2) $|f(x, y') - f(x, y'')| \leqslant L|y' - y''|$ for all points $(x, y')$, $(x, y'')$ in $D$, where $L$ is a given number;

(3) $d = \min\{|b - a|, c/M\}$, then there exists a unique

solution $y$ of the initial value problem given by Eqn (6.1), defined for $a \leqslant x \leqslant a + d$. ∎

*Example 6.1*

For the initial value problem
$$y^{(1)}(x) = 1 + x^2 + y^2, \quad y(0) = 0,$$
we have
$$f(x, y) = 1 + x^2 + y^2, \quad a = 0, \quad A = 0.$$

Taking $b = 1$, $c = 1$, $D$ is the rectangle $0 \leqslant x \leqslant 1$, $-1 \leqslant y \leqslant 1$. For $(x, y')$, $(x, y'')$ in $D$,
$$\begin{aligned}|f(x, y') - f(x, y'')| &= |y'^2 - y''^2| \\ &= |(y' + y'')(y' - y'')| \\ &\leqslant 2|y' - y''|,\end{aligned}$$
whence $L = 2$. Also for $(x, y)$ in $D$,
$$|f(x, y)| \leqslant 3,$$
so $M = 3$, and $d = \min\{1, \tfrac{1}{3}\} = \tfrac{1}{3}$. Hence the given initial value problem has a unique solution $y$ defined for $0 \leqslant x \leqslant \tfrac{1}{3}$ and $|y(x)| \leqslant 1$, $(0 \leqslant x \leqslant \tfrac{1}{3})$. These results do not imply that the initial value Problem 6.1 has no solution defined on a larger interval of $x$.

## 6.2 Difference Equations

In order to analyse the numerical methods to be considered subsequently we investigate briefly the solution of the inhomogeneous difference equation of order $n$ of the form
$$x_k + a_1 x_{k-1} + \ldots + a_n x_{k-n} = b, \tag{6.2}$$
in which $a_n \neq 0$, $b$ is a constant, and $k$ ranges over a set of consecutive integers. In order that the homogeneous equation
$$x_k + a_1 x_{k-1} + \ldots + a_n x_{k-n} = 0 \tag{6.3}$$
corresponding to Eqn (6.2) has a solution of the form
$$x_k = z^k, \tag{6.4}$$

the number $z$ must satisfy the equation

$$p_n(z) = z^n + a_1 z^{n-1} + \ldots + a_n = 0, \qquad (6.5)$$

obtained by substituting for $x_{k-j}$ ($j = 0, \ldots, n$) in Eqn (6.3) from Eqn (6.4).

A particular solution of Eqn (6.2) can be found by setting $x_k = c$ for all values of $k$ in Eqn (6.2), whence

$$c = \frac{b}{(1 + a_1 + \ldots + a_n)}. \qquad (6.6)$$

If $y_k$ satisfies Eqn (6.3) then clearly $y_k + c$ satisfies Eqn (6.2), and so if $\hat{z}$ satisfies Eqn (6.5) and $\alpha$ is a constant,

$$x_k = \alpha \hat{z}^k + \frac{b}{(1 + a_1 + \ldots + a_n)} \qquad (6.7)$$

satisfies Eqn (6.2). This solution is not the most general solution.

*Example 6.2*

Consider the difference equation

$$x_{k+1} + x_k = 2, \quad x_0 = 3, \quad (k = 0, 1, 2, \ldots).$$

For this equation,

$$p_n(z) = z + 1 = 0,$$

whence $\hat{z} = -1$. From Eqn (6.6) with $b = 2$, $a_1 = 1$, we obtain $c = 1$, whence a solution of the given difference equation is

$$x_k = \alpha(-1)^k + 1 \quad (k = 0, 1, 2, \ldots).$$

If $x_0 = 3$, we must have

$$3 = \alpha(-1)^0 + 1 = \alpha + 1,$$

whence $\alpha = 2$ and

$$x_k = 2(-1)^k + 1 \quad (k = 0, 1, 2, \ldots).$$

The theory of linear difference equations with constant coefficients is very extensive, but the preceding discussion together with a study of the Tutorial Examples 6.2 and 6.3 is sufficient for our purpose.

## 6.3 Taylor's Algorithm

Suppose that it is required to estimate the value of $y(x)$ on $[a, b]$ where $y$ satisfies Eqn (6.1). Let $N$ be a positive integer and let

$$h = (b - a)/N, \quad x_0 = a, \quad x_n = x_0 + nh \\ (n = 0, \ldots, N). \quad (6.8)$$

Then

$$y(x_{n+1}) = y(x_0 + (n+1)h) = y(x_n + h) \\ (n = 0, \ldots, N-1), \quad (6.9)$$

and using Taylor's theorem we obtain

$$y(x_{n+1}) = y(x_n) + \sum_{k=1}^{p} \frac{h^k}{k!} y^{(k)}(x_n) + \frac{h^{p+1}}{(p+1)!} y^{(p+1)}(\xi_n), \quad (6.10)$$

where $p$ is a positive integer, $x_n < \xi_n < x_{n+1}$, and $n = 0, \ldots, N-1$. Provided that $f$ is sufficiently differentiable with respect to $x$,

$$y^{(k)}(x) = f^{(k-1)}(x, y(x)), \quad (k = 1, 2, \ldots, p+1), \quad (6.11)$$

where

$$f^{(0)}(x, y) = f(x, y), \quad f^{(k)}(x, y) = \frac{d^k f(x, y)}{dx^k} \\ (k = 1, 2, \ldots, p+1).$$

Let

$$F_p(x, y; h) = \sum_{k=1}^{p} \frac{h^{k-1}}{k!} f^{(k-1)}(x, y). \quad (6.12)$$

Then since $f$ is known, $f^{(k)}(x, y)$ can be obtained in principle by repeatedly differentiating $f$. For example,

$$f^{(1)}(x, y) = \frac{\partial f}{\partial x} + f \frac{\partial f}{\partial y}, \quad (6.13)$$

$$f^{(2)}(x, y) = \frac{d}{dx} f^{(1)}(x, y) \\ = \frac{\partial^2 f}{\partial x^2} + 2f \frac{\partial^2 f}{\partial x \, \partial y} + f^2 \frac{\partial^2 f}{\partial y^2} + \frac{\partial f}{\partial y} \left[ \frac{\partial f}{\partial x} + f \frac{\partial f}{\partial y} \right]. \quad (6.14)$$

For higher derivatives the expressions rapidly become more complicated in general.

These formulae suggest the following algorithm for obtaining estimates $y_n$ of $y(x_n)$ ($n = 1, \ldots, N$).

*Algorithm 6.1* (Taylor's algorithm of order $p$)

With $h$ and $x_n$ ($n = 0, \ldots, N$) given by Eqn (6.8), set $y_0 = A$, and compute $y_n$ ($n = 1, \ldots, N$) from

$$y_{n+1} = y_n + h F_p(x_n, y_n; h) \quad (n = 0, \ldots, N-1).\blacksquare$$

Intuitively we would expect that as $h$ decreases, $y(x_n)$ is more closely approximated by $y_n$. Clearly $y_n$ is not in general equal to $y(x_n)$ because there is a truncation error $\varepsilon_n$ given by

$$\varepsilon_n = \frac{h^{p+1}}{(p+1)!} y^{(p+1)}(\xi_n) \quad (x_n < \xi_n < x_{n+1}), \quad (6.15)$$

and because the computed value of $h F_p(x_n, y_n; h)$ differs from the exact value due to rounding error. The truncation and rounding errors generated at each step accumulate as the computation proceeds, giving rise to overall or *cumulative* truncation and rounding errors. In general the cumulative truncation error decreases as $h$ decreases but the cumulative rounding error increases. This means that there is in general an optimum value of $h$ for which the combined cumulative truncation and rounding error is a minimum.

Taylor's algorithm of order 1 is particularly simple and is often called *Euler's method*. With $p = 1$ in Eqn (6.12), Euler's method consists of computing $y_n$ ($n = 1, \ldots, N$) from

$$y_{n+1} = y_n + h f(x_n, y_n) \quad (n = 0, \ldots, N-1). \quad (6.16)$$

*Example 6.3*

For the initial value problem

$$dy/dx = 3x^2 y, \quad y(0) = 1, \quad 0 \leqslant x \leqslant 1,$$

with $N = 5$, we have $a = 0$, $b = 1$, $h = 0.2$, $x_0 = 0$, $y_0 = y(0) = 1$, and $f(x, y) = 3x^2 y$. Hence from Eqn (6.16),

$$y_0 = 1\cdot0000,$$
$$y_1 = y_0 + hf(x_0, y_0) = 1\cdot0000,$$
$$y_2 = y_1 + hf(x_1, y_1) = 1\cdot0240,$$
$$y_3 = y_2 + hf(x_2, y_2) = 1\cdot1223,$$
$$y_4 = y_3 + hf(x_3, y_3) = 1\cdot3647,$$
$$y_5 = y_4 + hf(x_4, y_4) = 1\cdot8888.$$

The analytical solution of this initial value problem is
$$y(x) = \exp(x^3), \quad (0 \leqslant x \leqslant 1),$$
so
$$y(x_5) = \exp[(1\cdot0)^3] = 2\cdot7183,$$
whence the error in $y_5$ is $0\cdot8295$. To obtain more accurate estimates of $y(x)$ we must use very much smaller values of $h$, thereby increasing the computational labour. For this reason Euler's method is not often used. From Eqn (6.15) with $p = 1$ we see that the local truncation error $\varepsilon_n$ is $O(h^2)$.

Clearly for $p > 1$ in Taylor's algorithm we would intuitively expect the local truncation error and therefore also the cumulative truncation error to decrease more rapidly as $h$ decreases. However, the amount of computational labour per step increases rapidly as $p$ increases because in general the higher derivatives of $f$ are algebraically complicated; this also increases the rounding error per step or *local* rounding error. For example, in Taylor's algorithm of order 2, $\{y_n: n = 1, \ldots, N\}$ are computed from

$$y_0 = A, \quad y_{n+1} = y_n + h\bigg[f(x_n, y_n) + \frac{h}{2}\bigg\{\frac{\partial f(x_n, y_n)}{\partial x} + f(x_n, y_n)\frac{\partial f(x_n, y_n)}{\partial y}\bigg\}\bigg] \quad (6.17)$$

with $n = 0, \ldots, N-1$.

These considerations indicate that Taylor's algorithm can be rather inefficient.

## 6.4 Runge–Kutta Methods

In 1895 Runge suggested that it should be possible to obtain methods of essentially the same type as the Taylor methods

§6.4] RUNGE-KUTTA METHODS

without having to calculate the derivatives of $f$. This idea was investigated by Kutta, who published several methods in 1901 which were of a type now referred to as *Runge–Kutta Methods*. The general idea underlying these methods is as follows. Using the notation of Eqn (6.8) we compute $y_n$ ($n = 1, \ldots, N$) from

$$y_0 = A, \quad y_{n+1} = y_n + a_0 f_0 + \ldots + a_m f_m, \quad (6.18)$$

where

$$\begin{aligned}
f_0 &= hf(x_n, y_n), \\
f_1 &= hf(x_n + \alpha_1 h, y_n + \beta_{10} f_0), \\
f_2 &= hf(x_n + \alpha_2 h, y_n + \beta_{20} f_0 + \beta_{21} f_1), \\
&\ldots \\
f_m &= hf(x_n + \alpha_m h, y_n + \beta_{m0} f_0 + \ldots + \beta_{m,m-1} f_{m-1}).
\end{aligned} \quad (6.19)$$

In Eqns (6.18) and (6.19), $m \geqslant 1$ is an integer and the parameters $a_0, \ldots, a_m, \alpha_1, \ldots, \alpha_m$, and $\beta_{10}, \ldots, \beta_{m,m-1}$ are chosen in such a way that $y_n$ computed from Eqn (6.18) coincides with the result obtained using the Taylor algorithm of as high an order as possible. For $m > 2$, the algebra required to calculate the parameters becomes extremely complicated. The general method for obtaining the parameters will be illustrated for the simplest case, for which $m = 1$. Accordingly we seek $a_0, a_1, \alpha_1$, and $\beta_{10}$ such that $y_n$ ($n = 1, \ldots, N$) defined by

$$\begin{aligned}
y_0 &= A, \\
y_{n+1} &= y_n + a_0 f_0 + a_1 f_1 \quad (n = 0, \ldots, N-1), \\
f_0 &= hf(x_n, y_n), \\
f_1 &= hf(x_n + \alpha_1 h, y_n + \beta_{10} f_0),
\end{aligned} \quad (6.20)$$

will coincide with $y_n$ ($n = 1, \ldots, N$) obtained from Taylor's algorithm of as high an order as possible.

From Eqns (6.10), (6.12), (6.13), and (6.14) with $p = 3$, and using subscripts to denote partial differentiation with respect to $x$ and $y$, we obtain

$$y(x_{n+1}) = y(x_n) + hf + \frac{h^2}{2}(f_x + ff_y)$$
$$+ \frac{h^3}{6}(f_{xx} + 2ff_{xy} + f^2 f_{yy} + f_x f_y + ff_y^2) + \frac{h^4}{24} y^{(4)}(\xi_n)$$
$$(x_n < \xi_n < x_{n+1}) \quad (6.21)$$

in which all functions are to be evaluated at $x = x_n$, $y = y(x_n)$. To compare Eqns (6.20) and (6.21) we expand $f_1$ about $(x_n, y_n)$ by Taylor's theorem for functions of two variables, obtaining

$$f(x_n + \alpha_1 h, y_n + \beta_{10} f_0) = (f + \alpha_1 h f_x + \beta_{10} f_0 f_y$$
$$+ \alpha_1^2 \frac{h^2}{2} f_{xx} + \alpha_1 h \beta_{10} f_0 f_{xy} + \beta_{10}^2 f_0^2 f_{yy})$$
$$+ \left(\alpha_1 h \frac{\partial}{\partial x} + \beta_{10} f_0 \frac{\partial}{\partial y}\right)^3 f(\eta_n, \zeta_n), \quad (6.22)$$

where $x_n < \eta_n < x_n + \alpha_1 h$, $y_n < \zeta_n < y_n + \beta_{10} f_0$, and all functions are evaluated at $x = x_n$, $y = y_n$. Using Eqn (6.22) in Eqn (6.20) we obtain

$$y_{n+1} = y_n + (a_0 + a_1) hf + a_1 h^2 (\alpha_1 f_x + \beta_{10} f f_y)$$
$$+ a_1 h^3 \left(\frac{\alpha_1^2}{2} f_{xx} + \alpha_1 \beta_{10} f f_{xy} + \frac{\beta_{10}^2}{2} f^2 f_{xy}\right)$$
$$+ a_1 h^4 \left(\alpha_1 \frac{\partial}{\partial x} + \beta_{10} f \frac{\partial}{\partial y}\right)^3 f(\eta_n, \zeta_n). \quad (6.23)$$

Comparing Eqns (6.21) and (6.23) we see that to make the terms involving $h$ and $h^2$ of the same form we must have

$$\begin{aligned} a_0 + a_1 &= 1, \\ a_1 \alpha_1 &= \tfrac{1}{2}, \\ a_1 \beta_{10} &= \tfrac{1}{2}. \end{aligned} \quad (6.24)$$

It is clear that in general we cannot adjust the coefficients so as to make the terms involving $h^3$ of the same form. Since Eqns (6.24) are 3 equations for 4 unknowns, there are infinitely many possible choices for $a_0$, $a_1$, $\alpha_1$, and $\beta_{10}$. These choices give rise to the class of Runge–Kutta methods of order 2, which from Eqns (6.24) all have formulae of the form

$$y_0 = A, y_{n+1} = y_n + (1 - a_1) h f(x_n, y_n)$$
$$+ a_1 h f\left(x_n + \frac{h}{2a_1}, y_n + \frac{h}{2a_1} f(x_n, y_n)\right), \quad (6.25)$$

where $n = 0, \ldots, N - 1$.

Taking $a_1 = \frac{1}{2}$ in Eqn (6.25) we obtain Heun's method, in which for $n = 0, \ldots, N-1$,

$$y_{n+1} = y_n + \frac{h}{2}[f(x_n, y_n) + f\{x_n + h, y_n + hf(x_n, y_n)\}]. \quad (6.26)$$

Taking $a_1 = 1$ in Eqn (6.25) we obtain the modified Euler method, in which for $n = 0, \ldots, N-1$,

$$y_{n+1} = y_n + hf\left(x_n + \frac{h}{2}, y_n + \frac{h}{2}f(x_n, y_n)\right). \quad (6.27)$$

*Example 6.4*

For the initial value problem in Example 6.3 we compute $y_n$ ($n = 0, \ldots, N$) from

$$y_0 = 1, \quad y_{n+1} = y_n + 3h(x_n + h/2)^2(y_n + hx_ny_n)$$
$$(n = 0, \ldots, N-1)$$

if we use the modified Euler method, and from

$$y_0 = 1, \quad y_{n+1} = y_n + h\{3x_n^2 y_n + 3(x_n + h)^2(y_n + 2hx_ny_n)\}$$
$$(n = 0, \ldots, N-1)$$

if we use Heun's method.

To obtain an expression for the local truncation error $\varepsilon_n$ for Runge–Kutta methods of order 2, we note that

$$y(x_{n+1}) = y(x_n) + (1 - a_1)hf + a_1h\{f + \frac{h}{2a_1}f_x$$
$$+ \frac{h}{2a_1}ff_y + \frac{1}{2}\left(\frac{h}{2a_1}\frac{\partial}{\partial x} + \frac{h}{2a_1}f\frac{\partial}{\partial y}\right)^2 f(\eta_n, \zeta_n)\} + \varepsilon_n, \quad (6.28)$$

where $f, f_x$, and $f_y$ are to be evaluated at $x = x_n, y = y(x_n)$ and $x_n < \eta_n < x_{n+1}$, $y(x_n) < \zeta_n < y(x_{n+1})$. Also we have by Taylor's theorem

$$y(x_{n+1}) = y(x_n) + hf + \frac{h^2}{2}(f_x + ff_y) + \frac{h^3}{6}y^{(3)}(\xi_n), (6.29)$$

where $f, f_x$, and $f_y$ are to be evaluated at $x = x_n, y = y(x_n)$, and $x_n < \xi_n < x_{n+1}$.

Comparing Eqns (6.28) and (6.29) we obtain

$$\varepsilon_n = h^3\left\{\frac{1}{6}y^{(3)}(\xi_n) - \frac{1}{8a_1}\left(\frac{\partial}{\partial x} + f\frac{\partial}{\partial y}\right)^2 f(\eta_n, \xi_n)\right\}. \quad (6.30)$$

Clearly if $f$ and the requisite partial derivatives of $f$ are bounded in a sufficiently large rectangle then there is a number $M$ such that

$$\varepsilon_n \leqslant Mh^3. \quad (6.31)$$

Hence for Runge–Kutta methods of order 2 the local truncation error is $O(h^3)$ as it is for Taylor's algorithm of order 2.

Higher-order Runge–Kutta methods are available. With $m = 2$ in Eqn (6.18) there are 6 equations connecting the 8 parameters $a_0$, $a_1$, $a_2$, $\alpha_1$, $\alpha_2$, $\beta_{10}$, $\beta_{20}$, and $\beta_{21}$, and the choices are doubly infinite. The following formulae correspond to two of the commonly used choices of the parameters.

(1) (Kutta)

$$y_{n+1} = y_n + (f_0 + 4f_1 + f_2)/6 \quad (n = 0, \ldots, N-1)$$
$$f_0 = hf(x_n, y_n),$$
$$f_1 = hf\left(x_n + \frac{h}{2}, y_n + \frac{f_0}{2}\right),$$
$$f_2 = hf(x_n + h, y_n + 2f_1 - f_0),$$
$$\varepsilon_n = O(h^4).$$

(2) (Heun)

$$y_{n+1} = y_n + (f_0 + 3f_2)/4 \quad (n = 0, \ldots, N-1)$$
$$f_0 = hf(x_n, y_n),$$
$$f_1 = hf\left(x_n + \frac{h}{3}, y_n + \frac{f_0}{3}\right),$$
$$f_2 = hf\left(x_n + \frac{2h}{3}, y_n + \frac{2f_1}{3}\right),$$
$$\varepsilon_n = O(h^4).$$

Still higher-order Runge–Kutta methods are available. The following method is associated with Runge, Kutta, and Simpson.

§6.4] RUNGE–KUTTA METHODS

$y_{n+1} = y_n + (f_0 + 2f_1 + 2f_2 + f_3)/6 \quad (n = 0, \ldots, N-1),$
$f_0 = hf(x_n, y_n),$
$f_1 = hf\left(x_n + \frac{h}{2}, y_n + \frac{f_0}{2}\right),$
$f_2 = hf\left(x_n + \frac{h}{2}, y_n + \frac{f_1}{2}\right),$
$f_3 = hf(x_n + h, y_n + f_2),$
$\varepsilon_n = O(h^5).$

*Example 6.5*

If we denote the combined cumulative truncation and rounding error in the estimate $y_n$ of $y(x_n)$ by $e(x_n)$, so that

$$e(x_n) = y(x_n) - y_n \quad (n = 0, \ldots, N) \qquad (6.32)$$

then for the initial value problem in Example 6.3, using the Runge–Kutta method (1) we obtain the results shown in Table 6.1.

|       |          | $h = 0.2$ |                              | $h = 0.1$ |                              | $h = 0.01$ |                              |
|-------|----------|-----------|------------------------------|-----------|------------------------------|------------|------------------------------|
| $x_n$ | $y(x_n)$ | $y_n$     | $\|e(x_n)\| \times 10^3$ | $y_n$     | $\|e(x_n)\| \times 10^3$ | $y_n$      | $\|e(x_n)\| \times 10^5$ |
| 0.0   | 1·000000 | 1·000000  | 0·000                        | 1·000000  | 0·000                        | 1·000000   | 0·000                        |
| 0.2   | 1·008032 | 1·008048  | 0·016                        | 1·008036  | 0·004                        | 1·008032   | 0·002                        |
| 0.4   | 1·066092 | 1·066310  | 0·218                        | 1·066124  | 0·032                        | 1·066092   | 0·013                        |
| 0.6   | 1·241102 | 1·241891  | 0·789                        | 1·241216  | 0·114                        | 1·241102   | 0·044                        |
| 0.8   | 1·668625 | 1·670392  | 1·767                        | 1·668898  | 0·273                        | 1·668624   | 0·142                        |
| 1.0   | 2·718282 | 2·718844  | 0·562                        | 2·718442  | 0·160                        | 2·718277   | 0·502                        |

TABLE 6.1

## 6.5 Analysis of Single-Step Methods

Integrating Eqn (6.1) over $[x_n, x_{n+1}]$ we obtain

$$y(x_{n+1}) = y(x_n) + \int_{x_n}^{x_{n+1}} f(t, y(t)) \, dt. \tag{6.33}$$

Approximating the integral in Eqn (6.33) by $hF(x_n, y(x_n); h)$, we obtain

$$\int_{x_n}^{x_{n+1}} f(t, y(t)) \, dt = hF(x_n, y(x_n); h) + \varepsilon_n, \tag{6.34}$$

where $\varepsilon_n$ is the truncation error. Using Eqn (6.34) in Eqn (6.33) we obtain

$$y(x_{n+1}) = y(x_n) + hF(x_n, y(x_n); h) + \varepsilon_n. \tag{6.35}$$

By neglecting the local truncation error $\varepsilon_n$ we obtain the general *explicit single-step method*

$$y_0 = A, \quad y_{n+1} = y_n + hF(x_n, y_n; h) \quad (n = 0, \ldots, N-1). \tag{6.36}$$

Taylor's algorithm and the Runge–Kutta methods are all single-step methods for the numerical solution of Eqn (6.1).

Let $F$ satisfy the Lipschitz condition

$$|F(x, y'; h) - F(x, y''; h)| \leqslant L|y' - y''| \tag{6.37}$$

for all $(x, y')$, $(x, y'')$ in the rectangle $R$ consisting of all points $(x, y)$ such that $a \leqslant x \leqslant a + d$, $A - r \leqslant y \leqslant A + r$, where $r > c$ and the notation of Theorem 6.1 has been used. Suppose also that there is a number $M$ such that

$$|\varepsilon_n| \leqslant Mh^{p+1} \quad (n = 0, \ldots, N-1). \tag{6.38}$$

With

$$e_n = y(x_n) - y_n \quad (n = 0, \ldots, N) \tag{6.39}$$

we obtain, on subtracting Eqn (6.36) from Eqn (6.33), assuming that $(x_k, y_k)$ is in $R$ for $k = 0, \ldots, n$, and using Eqns (6.37) and (6.38)

$$|e_{n+1}| \leqslant (1 + hL)|e_n| + Mh^{p+1}. \tag{6.40}$$

Also since $y_0 \,(=A)$ is given, $e_0 = 0$.
Define the sequence $\{u_k\}$ by
$$u_0 = 0, \quad u_{k+1} = (1 + hL)u_k + Mh^{p+1} \quad (k = 0, \ldots, n). \quad (6.41)$$
Then it is easy to show by induction that
$$|e_k| \leqslant u_k \quad (k = 0, \ldots, n). \quad (6.42)$$
From Section 6.2 we see that the difference equation (6.41) has the solution
$$u_k = \frac{Mh^p}{L}[(1 + hL)^k - 1] \quad (k = 0, \ldots, n). \quad (6.43)$$
Also since $hL > 0$, and
$$\exp(khL) = 1 + khL + \sum_{s=2}^{\infty} \frac{(khL)^s}{s!} \quad (0 \leqslant hL < \infty),$$
we have
$$(1 + khL) \leqslant \exp(khL), \quad (k = 0, 1, 2, \ldots) \quad (6.44)$$
whence, by inequality (6.42) we have
$$|e_k| \leqslant \frac{Mh^p}{L}[\exp\{(x_k - x_0)L\} - 1] \quad (k = 0, \ldots, n). \quad (6.45)$$

Next we show that $(x_{n+1}, y_{n+1})$ is in $R$. We have
$$|A - y_{n+1}| = |A - y(x_{n+1}) + e_{n+1}| \leqslant |A - y(x_{n+1})| + |e_{n+1}|.$$

Using inequality (6.40) and recalling from Theorem 6.1 that $|A - y(x_{n+1})| < c$ we obtain
$$|A - y_{n+1}| \leqslant c + (1 + hL)|e_n| + Mh^{p+1}. \quad (6.46)$$

Using inequality (6.45) in inequality (6.46) it is easy to see that if $h$ is sufficiently small, then $|A - y_{n+1}| \leqslant r$ and $(x_{n+1}, y_{n+1})$ is in $R$. Hence by induction $(x_n, y_n)$ is in $R$ for $n = 0, \ldots, N$ and we have, for the cumulative truncation error $e_n$, the bound
$$|e_n| \leqslant \frac{Mh^p}{L}[\exp\{(x_n - x_0)L\} - 1] \quad (n = 0, \ldots, N). \quad (6.47)$$

Hence if we decrease $h$ and increase $n$ so as to keep $nh$ and therefore $x_n$ constant, we see that $|e_n| \to 0$, and that there is a number $K$ such that
$$|e_n| \leqslant Kh^p \quad (n = 0, \ldots, N). \quad (6.48)$$

Hence we see that in general if the local truncation error is $O(h^{p+1})$ then the cumulative truncation error is $O(h^p)$. A more detailed analysis shows that the combined truncation and rounding error $e(x_n)$ is bounded by

$$|e(x_n)| \leqslant \left(\frac{Mh^p}{L} + \frac{\Delta_N}{hL}\right)[\exp\{(x_n - x_0)L\} - 1] + |\delta_0| \exp\{(x_n - x_0)L\} \quad (6.49)$$

where $n = 0, \ldots, N$, $\delta_0$ is the rounding error in $y_0$, and if $\rho_k$ is the rounding error arising from the computation of $y_{k+1}$ from $y_k$, then

$$\Delta_N = \max_{0 \leqslant k \leqslant N} \{|\rho_k|\}.$$

The term in $\Delta_N/hL$ bounds the cumulative rounding error due to the local rounding errors $\rho_n$ ($n = 0, \ldots, N$), and increases as $h$ decreases, giving rise to an optimum value of $h$ as explained previously. The term in $|\delta_0|$ bounds the cumulative rounding error due to the rounding error in $y_0$, and is important because it suggests that we cannot expect to estimate $y(x_n)$ with greater accuracy than that of $y_0$.

*Example 6.6*

For Taylor's algorithm of order $p$, $F$ ($= F_p$) is given by Eqn (6.12), and $\varepsilon_n$ is given by Eqn (6.15). Hence we see that for Taylor's algorithm of order $p$, the cumulative truncation error $e_n$ is $O(h^p)$.

For the Runge–Kutta methods of order 2, we have from Eqn (6.25),

$$F(x, y; h) = (1 - a_1)f(x, y) + a_1 f\left(x + \frac{h}{2a_1}, y + \frac{h}{2a_1}f(x, y)\right),$$

and $\varepsilon_n$ is given by Eqn (6.30) and bounded according to inequality (6.31). Hence $e_n$ is $O(h^2)$. In general, a single-step method for which $e_n$ is $O(h^p)$ is called a single-step method *of order p*.

## 6.6 Adams–Bashforth Methods

In Eqn (6.33) let us approximate $f(t, y(t))$ on $[x_n, x_{n+1}]$ by the interpolating polynomial $p_q(t)$ of degree at most $q$ which interpolates $f$ on $\{x_n, \ldots, x_{n-q}\}$, where $n \geqslant q$. Using Newton's backward difference form of the interpolating polynomial we have

$$p_q(t) = \sum_{k=0}^{q} \binom{u+k-1}{k} \nabla^k f_n,$$

where

$$u = \frac{(t - x_n)}{h}, \quad f_n = f(x_n, y(x_n)).$$

Then if

$$\int_{x_n}^{x_{n+1}} f(t, y(t)) \, dt = \int_{x_n}^{x_{n+1}} p_q(t) \, dt + \varepsilon_n,$$

and

$$\gamma_k = \int_0^1 \binom{u+k-1}{u} du \quad (k = 0, \ldots, q), \tag{6.50}$$

we have from Eqn (6.33)

$$y(x_{n+1}) = y(x_n) + h \sum_{k=0}^{q} \gamma_k \nabla^k f_n + \varepsilon_n \quad (n = q, \ldots, N-1).$$

The Adams–Bashforth methods then consist of computing $\{y_n : n = q+1, \ldots, N\}$ from

$$y_{n+1} = y_n + h \sum_{k=0}^{q} \gamma_k \nabla^k f_n \quad (n = q, \ldots, N-1), \tag{6.51}$$

given the $(q+1)$ starting values $\{y_n : n = 0, \ldots, q\}$. It is often more convenient to express the $\nabla^k f_n$ ($k = 0, \ldots, q$) in terms of $f_k$ ($k = 0, \ldots, q$). A frequently used Adams–Bashforth method corresponds to $q = 3$. From Eqn (6.50), we have

$$\gamma_0 = 1, \quad \gamma_1 = \tfrac{1}{2}, \quad \gamma_2 = \tfrac{5}{12}, \quad \gamma_3 = \tfrac{3}{8}, \quad \gamma_4 = \tfrac{251}{720}.$$

Using these and Example 2.7 we obtain from Eqn (6.51)

$$y_{n+1} = y_n + \frac{h}{24}[55f_n - 59f_{n-1} + 37f_{n-2} - 9f_{n-3}]$$
$$(n = 3, \ldots, N - 1) \quad (6.52)$$

Using Eqn (2.38) and noting that $\binom{u+3}{4}$ does not change sign for $0 \leq u \leq 1$, we obtain also

$$\varepsilon_n = \int_{x_n}^{x_{n+1}} [f(t, y(t)) - p_q(t)] \, dt = \tfrac{251}{720} h^5 f^{(4)}(\eta, y(\eta)), \quad (6.53)$$

where $x_{n-3} < \eta < x_{n+1}$. Hence $\varepsilon_n$ is $O(h^5)$. Analysis similar to that of Section 6.5 shows that if the starting values $\{y_k : k = 0, \ldots, q\}$ for the general Adams–Bashforth method have $e_k = O(h^{q+1})$, then $e_n = O(h^{q+1})$ $(n = 0, \ldots, N)$. Hence for $q = 3$ we need starting values with errors $O(h^4)$ to obtain a cumulative truncation error $O(h^4)$. The Runge–Kutta–Simpson method of Section 6.4 with $\varepsilon_n = O(h^5)$ could be used to compute starting values, $y_k$ ($k = 1, 2, 3$) for Eqn (6.52).

*Example 6.7*

Applying Eqn (6.52) to the initial value problem of Example 6.3 we compute $\{y_n : n = 4, \ldots, N\}$ from

$$y_{n+1} = y_n + \frac{h}{8}[55x_n^2 y_n - 59x_{n-1}^2 y_{n-1} + 37x_{n-2}^2 y_{n-2} - 9x_{n-3}^2 y_{n-3}]$$

where $n = 3, \ldots, N - 1$. Using exact starting values we obtain the results shown in Table 6.2.

Note that for the Adams–Bashforth fourth-order method corresponding to Eqn (6.52) only *one* evaluation of $f$ is needed for each step, compared with *four* evaluations of $f$ in the Runge–Kutta methods of order four. This motivates the use of Adams–Bashforth methods with starting values computed from appropriate Runge–Kutta methods.

Formulae for $y_{n+1}$ in terms of $y_{n-k}$ ($k = 0, \ldots, q$) are called *multistep formulae*. Many multistep methods exist. The general

|       |          | $h = 0.2$ |                             | $h = 0.1$ |                             | $h = 0.01$ |                             |
|-------|----------|----------|-----------------------------|----------|-----------------------------|-----------|-----------------------------|
| $x_n$ | $y(x_n)$ | $y_n$    | $|e(x_n)| \times 10^1$ | $y_n$    | $|e(x_n)| \times 10^2$ | $y_n$     | $|e(x_n)| \times 10^3$ |
| 0.0   | 1·000000 | 1·000000 | 0·000 | 1·000000 | 0·000 | 1·000000 | 0·000 |
| 0.2   | 1·008032 | 1·008032 | 0·000 | 1·008032 | 0·000 | 1·008024 | 0·008 |
| 0.4   | 1·066092 | 1·066092 | 0·000 | 1·065831 | 0·026 | 1·066075 | 0·017 |
| 0.6   | 1·241102 | 1·241102 | 0·000 | 1·239493 | 0·161 | 1·241070 | 0·032 |
| 0.8   | 1·668625 | 1·641143 | 0·275 | 1·662241 | 0·638 | 1·668568 | 0·057 |
| 1.0   | 2·718282 | 2·575032 | 1·432 | 2·691128 | 2·715 | 2·718166 | 0·116 |

TABLE 6.2

multistep method can be analysed in much the same way as for the general single-step method. It is found that if the local truncation error is $O(h^{p+1})$ and the cumulative truncation error in the starting values is $O(h^p)$, then the cumulative truncation error for the method is $O(h^p)$. The behaviour of the rounding error is similar to that for single-step methods. Some multistep formulae are not computationally satisfactory because of rapid error growth due to the phenomenon of numerical instability, further consideration of which is beyond the scope of this book.

## 6.7 Adams–Moulton Methods

In Eqn (6.33) let us approximate $f(t, y(t))$ on $[x_n, x_{n+1}]$ by the interpolating polynomial $p_{q+1}(t)$ of degree at most $q + 1$ which interpolates $f$ on $\{x_{n+1}, \ldots, x_{n-q}\}$ where $n \geq q$. Using Newton's backward difference form of the interpolating polynomial we have

$$p_{q+1}(t) = \sum_{k=0}^{q+1} \binom{u + k - 1}{k} \nabla^k f_{n+1},$$

where

$$u = \frac{(t - x_{n+1})}{h}, \quad f_{n+1} = f(x_{n+1}, y(x_{n+1})).$$

Then with

$$\hat{\gamma}_k = \int_{-1}^{0} \binom{u+k-1}{k} du \quad (k = 0, \ldots, q+1), \quad (6.54)$$

and

$$\int_{x_n}^{x_{n+1}} f(t, y(t)) \, dt = \int_{x_n}^{x_{n+1}} p_{q+1}(t) \, dt + \hat{\varepsilon}_n,$$

we have from Eqn (6.33)

$$y(x_{n+1}) = y(x_n) + h \sum_{k=0}^{q+1} \hat{\gamma}_k \nabla^k f_{n+1} + \hat{\varepsilon}_n \quad (n = q, \ldots, N-1).$$

The Adams–Moulton formulae are obtained by neglecting $\hat{\varepsilon}_n$ giving

$$y_{n+1} = y_n + h \sum_{k=0}^{q+1} \hat{\gamma}_k \nabla^k f_{n+1} \quad (n = q, \ldots, N-1). \quad (6.55)$$

Unlike the Adams–Bashforth formulae which are *explicit* because $y_{n+1}$ does not occur on the right-hand side, the Adams–Moulton formulae are *implicit* because $\nabla^k f_{n+1}$ is a function of $y_{n+1}$ for $k = 0, \ldots, q+1$. This presents no difficulty in practice as we shall see.

With $q = 2$ in Eqns (6.54) and (6.55) and using Example 2.7 we obtain the frequently used Adams–Moulton formula

$$y_{n+1} = y_n + (h/24)[9f_{n+1} + 19f_n - 5f_{n-1} + f_{n-2}]. \quad (6.56)$$

Using Eqn (2.38) and noting that $\binom{u+3}{4}$ does not change sign for $-1 \leqslant u \leqslant 0$ we obtain for this formula

$$\hat{\varepsilon}_n = \int_{x_n}^{x_{n+1}} [f(t, y(t)) - p_{q+1}(t)] \, dt$$
$$= -\tfrac{19}{720} h^5 f^{(4)}(\hat{\eta}, y(\hat{\eta})), \quad (6.57)$$

where $x_{n-2} < \hat{\eta} < x_{n+1}$. Hence $\hat{\varepsilon}_n$ is $O(h^5)$.

If $y_0, \ldots, y_n$ are known, Eqn (6.56) can be solved iteratively for $y_{n+1}$ by generating the sequence $\{y_{n+1,i}\}$ from

$$y_{n+1,i+1} = y_n + (h/24)[9f(x_{n+1}, y_{n+1,i}) + 19f_n - 5f_{n-1} + f_{n-2}]$$

for $i = 0, 1, 2, \ldots$, if an initial estimate $y_{n+1,0}$ is known. Applying Theorem 4.3 we see that the iterative sequence converges to $y_{n+1}$ if $y_{n+1,0}$ is sufficiently close to $y_{n+1}$ and if

$$M \cdot \frac{9h}{24} < 1, \quad \left( M = \max \left| \frac{\partial f}{\partial y} \right| \right).$$

A suitable method for obtaining $y_{n+1,0}$ is to use the Adams–Bashforth formula (6.52). It can then be shown that if $h$ is sufficiently small, only one iteration is required to obtain a sufficiently accurate estimate of $y_{n+1}$. When used in this way, Eqn (6.52) is called a *predictor* and Eqn (6.56) is called a *corrector*. Many other predictor–corrector pairs of formulae exist. An analysis of the general predictor–corrector method shows that if the cumulative truncation error of the starting values is $O(h^p)$ and the local truncation errors of the predictor and corrector are $O(h^p)$ and $O(h^{p+1})$ respectively then under fairly general conditions the cumulative truncation error of the method is $O(h^p)$. Again as for multistep methods the behaviour of the cumulative rounding error is similar to that for single-step methods.

Combining Eqns (6.52) and (6.56) we obtain the fourth-order Adams–Moulton predictor–corrector method given in the following algorithm.

*Algorithm 6.2*

Given $\{y_k: k = 0, \ldots, 3\}$ and using the notation of Eqn (6.8) compute $\hat{y}_{n+1}$ and $y_{n+1}$ for $n = 3, \ldots, N-1$ from

$$\hat{y}_{n+1} = y_n + (h/24)[55f_n - 59f_{n-1} + 37f_{n-2} - 9f_{n-3}],$$

where

$$f_{n-k} = f(x_{n-k}, y_{n-k}) \quad (k = 0, 1, 2, 3),$$

and

$$y_{n+1} = y_n + (h/24)[9f(x_{n+1}, \hat{y}_{n+1}) + 19f_n - 5f_{n-1} + f_{n-2}]. \blacksquare$$

*Example 6.8*

Applying Algorithm 6.2 to the initial value problem of Example 6.3 we compute $\{y_n: n = 4, \ldots, N\}$ from

$$\hat{y}_{n+1} = y_n + (h/8)[55x_n^2 y_n - 59x_{n-1}^2 y_{n-1} + 37x_{n-2}^2 y_{n-2} - 9x_{n-3}^2 y_{n-3}],$$

$$y_{n+1} = y_n + (h/8)[9x_{n+1}^2 \hat{y}_{n+1} + 19x_n^2 y_n - 5x_{n-1}^2 y_{n-1} + x_{n-2}^2 y_{n-2}].$$

Using the exact starting values we obtain the results shown in Table 6.3.

|       |          | $h = 0.2$ |                            | $h = 0.1$ |                            | $h = 0.01$ |                            |
|-------|----------|-----------|----------------------------|-----------|----------------------------|------------|----------------------------|
| $x_n$ | $y(x_n)$ | $y_n$     | $\|e(x_n)\| \times 10^3$   | $y_n$     | $\|e(x_n)\| \times 10^3$   | $y_n$      | $\|e(x_n)\| \times 10^3$   |
| 0·0   | 1·000000 | 1·000000  | 0·000                      | 1·000000  | 0·000                      | 1·000000   | 0·000                      |
| 0·2   | 1·008032 | 1·008032  | 0·000                      | 1·008032  | 0·000                      | 1·008024   | 0·008                      |
| 0·4   | 1·066092 | 1·066092  | 0·000                      | 1·066119  | 0·027                      | 1·066074   | 0·018                      |
| 0·6   | 1·241102 | 1·241102  | 0·000                      | 1·241229  | 0·127                      | 1·241070   | 0·032                      |
| 0·8   | 1·668625 | 1·669336  | 0·711                      | 1·669011  | 0·386                      | 1·668570   | 0·055                      |
| 1·0   | 2·718282 | 2·716536  | 1·746                      | 2·719284  | 1·002                      | 2·718177   | 0·105                      |

TABLE 6.3

## Tutorial Examples

1. Use Theorem 6.1 to show that the initial value problem
$$dy/dx = 3x^2 y, \quad y(0) = 1$$
has a unique solution for $0 \leqslant x \leqslant 1/6$.

2. Show that a solution of the difference equation
$$x_{k+1} + ax_k = b \quad (k = 0, 1, 2, \ldots), \, x_0 = c$$
is
$$x_k = \left[c - \frac{b}{(1+a)}\right](-a)^k + \frac{b}{(1+a)} \quad (k = 0, 1, 2, \ldots),$$

if $a \neq -1$. Let $x_k$ and $y_k$ be solutions of the given difference equation with $x_0 = c$, $y_0 = c$. Write down a difference equation which is satisfied by $(x_k - y_k)$ and solve it. Hence show that the given solution of the original difference equation is unique. If $a = -1$, obtain a solution of the given difference equation in the form

$$x_k = pk + q \quad (k = 0, 1, 2, \ldots).$$

3. Obtain a solution of the difference equation

$$x_{k+2} + 3x_{k+1} + 2x_k = 1, \quad x_0 = 0.$$

4. Using Theorem 6.1 show that the initial value problem

$$dy/dx = 1 + y^2, \quad y(0) = 0$$

has a unique solution for $0 \leqslant x \leqslant 0.5$.

With $h = 0.1$, estimate the value of $y(0.5)$ using Euler's method and Taylor's algorithm of orders 2 and 3. Repeat the computation with $h = 0.05$.

5. For the initial value problem in Tutorial Example 6.4 estimate the value of $y(0.5)$ using the Runge–Kutta methods in Section 6.4 with $h = 0.1$ and $h = 0.05$.

6. Write down the Adams–Bashforth formula corresponding to $q = 1$ in Eqn (6.51) and obtain an expression for its local truncation error.

Write down a suitable Adams–Moulton formula for use as a corrector in conjunction with the Adams–Bashforth formula corresponding to $q = 1$. Select a method for obtaining starting values of suitable accuracy and hence estimate $y(0.5)$ for the initial value problem in Tutorial Example 6.4 with $h = 0.1$ and $h = 0.05$. Estimate $y(0.5)$ using the fourth-order Adams–Moulton method with $h = 0.05$ and the appropriate starting values which were computed in Tutorial Example 6.5.

# Answers to Tutorial Examples

### Chapter 1

4. Set $b_n = a_n$ and compute $\{b_k: k = n-1, \ldots, 0\}$ from
$$b_{n-k} = a_{n-k} + \alpha b_{n-k+1} \quad (k = 1, \ldots, n).$$

Set $p_n(\alpha) = b_0$.

Set $c_n = b_n$ and compute $\{c_k: k = n-1, \ldots, 1\}$ from
$$c_{n-k} = b_{n-k} + \alpha c_{n-k+1} \quad (k = 1, 2, \ldots, n-1).$$

Set $p_n^{(1)}(\alpha) = c_1$. ∎

To compute $p_n(\alpha)$ using this algorithm we need $n$ multiplications and $n$ additions, and to compute $p_n^{(1)}(\alpha)$ we need $(n-1)$ multiplications and $(n-1)$ additions.

To compute $p_n(\alpha)$ by forming each term separately we need $n(n+1)/2$ multiplications and $n$ additions. To compute $p_n^{(1)}(\alpha)$ by forming each term separately we need $n(n-1)/2$ multiplications and $(n-1)$ additions.

5. One significant figure.

7. (a) $\alpha/[(x+\alpha)^{1/2} + x^{1/2}]$;

   (b) $\dfrac{2}{\alpha} \sin\left(\dfrac{\alpha}{2}\right) \cos\left(x + \dfrac{\alpha}{2}\right)$;

   (c) $-\alpha/[x(x+\alpha)]$;

   (d) $\log\left(1 + \dfrac{\alpha}{x}\right)$;

   (e) $-2\sin\left(x + \dfrac{\alpha}{2}\right) \sin\left(\dfrac{\alpha}{2}\right)$.

### Chapter 2

1. Yes.

ANSWERS TO TUTORIAL EXAMPLES

2. By Theorem 2.1 we conclude immediately that
$$p_3(x) = x^2 + 2x + 1 \quad (0 \leqslant x \leqslant 3).$$

4. 0·01.

5. Let $x - x_0 = u$. Then $h \leqslant u \leqslant 2h$, and on this interval,
$|L_0(x)| < 1/3$, $|L_1(x)| < 2$, $|L_2(x)| < 2$, $|L_3(x)| < 1/3$,
whence $|e_k(x)| < 14\varepsilon/3$.

6. $p_2(x) = 1\cdot8150x^2 - 3\cdot4915x + 3\cdot3499$.
sinh $(1\cdot95) = 3\cdot4430$, correct to 4D.
$|e_T(x)| \leqslant (x^3 - 6\cdot00x^2 + 11\cdot99x - 7\cdot98) \cosh(2\cdot1)/6$,
$x = 1\cdot95$.
$$|e_R(x)| \leqslant \left( \sum_{k=0}^{2} |L_k(x)| \right) \times 10^{-5}/2, \quad x = 1\cdot95.$$

7. $S(6) = 441$.

9. The errors in $\Delta^5 f_0$ and $\Delta^5 f_1$ due to an error $\varepsilon$ in $f_3$ are $10\varepsilon$ and $-10\varepsilon$ respectively. The erroneous values of $\Delta^5 f_0$ and $\Delta^5 f_1$ are 220 and 20 respectively. Hence $\varepsilon = 10$.

10. sinh $1\cdot85 = 3\cdot10129$, correct to 5D.
$$|e_T(x)| \leqslant \binom{s}{5} h^5 \cosh(2\cdot2), \quad s = \tfrac{1}{2}, \quad h = 0\cdot1.$$

11. sinh $2\cdot15 = 4\cdot23420$, correct to 5D.
$$|e_T(x)| \leqslant \binom{s+4}{5} h^5 \cosh(2\cdot2), \quad s = -\tfrac{1}{2}, \quad h = 0\cdot1.$$

The rounding error $e_R(s)$ in Newton's backward difference formula due to rounding errors $\varepsilon_k$ in $f_k$ ($k = 0, \ldots, n$) is bounded according to
$$|e_R(s)| < \sum_{k=0}^{n} \sum_{i=0}^{k} \left| \frac{(s+k-1)(s+k-2)\ldots s}{i! \ (k-i)!} \right| \cdot |\varepsilon_{n-i}|.$$
In this take
$n = 4$, $s = -\tfrac{1}{2}$, $\varepsilon_m = 10^{-5}/2$ $(m = 0, \ldots, 4)$.

## Chapter 3

1. Using the first formula with $f_0 = 0.295520$ and $h = 0.001$, we obtain $\cos 0.302 = 0.955083$. Using the second formula with $f_0 = 0.295520$ and $h = 0.002$, we obtain $\cos 0.302 = 0.954750$. Note that this result is more accurate than the first.

2. $f_0^{(1)} \approx (1/2h)[3f_0 - 4f_{-1} + f_{-2}]$.

With $h = 0.001$, we obtain $\cos 0.304 = 0.9535$, and with $h = 0.002$ we obtain $\cos 0.304 = 0.9542$. Note that the latter result is more accurate than the former even though the value of $h$ used for the latter is greater than for the former.

3. $|E_{T1}| \leqslant \dfrac{0.299339}{30} h^4 < \dfrac{h^4}{100}$;

$|E_{T2}| \leqslant \dfrac{0.299339}{6} h^2 < \dfrac{h^2}{30}$.

Hence for $h < 1$, the bound on the truncation error for the first formula is much less than for the second.

The structure of the first formula shows that it could be subject to large rounding error due to the factors $\pm 8/12$ and the fact that there are 4 values of $f$ each of which contributes its rounding error.

4. $|E_{T1}| \leqslant \dfrac{h^4}{90}[|f^{(5)}(\eta_1)| + 4|f^{(5)}(\eta_2)|]$. $(x_0 < \eta_1, \eta_2 < x_4)$;

$|E_{T2}| \leqslant \dfrac{h^2}{6}[4|f^{(3)}(\eta_2)| + 2|f^{(3)}(\eta_1)|]$. $(x_{-2} < \eta_1, \eta_2 < x_0)$.

5. $|E_{R1}| \leqslant 3\varepsilon/2h$,
$|E_{R2}| \leqslant \varepsilon/h$.

With $\varepsilon = 0.5 \times 10^{-6}$ and $h = 0.001$, $|E_{R1}| \leqslant 0.75 \times 10^{-3}$, and $|E_{R2}| = 2|E_{R1}|/3$. Hence the *bounds* on the *total* errors are much the same. The considerable difference in the *actual* errors is due to loss of significance resulting from the cancellation of very nearly equal numbers in the first formula.

## Chapter 4

1. (1) $L = 5/4$; (2) $L = 2$; (3) $L = 3$; (4) $L = 1$.

2. (1) inapplicable; (2) inapplicable; (3) inapplicable; (4) applicable, with $L = 1/2$.

3. We need 10 evaluations of $f$; $x^* = 0.3333$ correct to 4D.

4. Generate $\{x_n\}$ from $x_{n+1} = \exp(x_n - 2)$ $(n = 0, 1, 2, \ldots)$; this sequence converges if $0.1 \leqslant x_0 \leqslant 0.2$. With $x_0 = 0.1$, convergence to 6D is obtained with $n = 7$. For the bisection method, 19 evaluations of $f$ are needed.

5. The sequence $\{x_n\}$ generated from $x_0 = 0.42$,
$$x_{n+1} = \tfrac{1}{2}\cos^2 x_n \quad (n = 0, 1, 2, \ldots)$$
converges to $x^*$ with an accuracy of at least 6D in 14 iterations.

7. The value of the limit is $1/2$.

8. $x^* = 0.3333$ correct to 4D in 2 iterations.

9. For the secant method $x^* = 0.3333333$ correct to 7D in 4 iterations. For the method of false position, $x^* = 0.3333333$ correct to 7D in 7 iterations.

## Chapter 6

3. $x_k = \alpha[(-1)^k - (-2)^k] + (1/6)[1 - (-2)^k]$.

6. The Adams–Bashforth formula for $q = 1$ in Eqn (6.51) is
$$y_{n+1} = y_n + h/2[3f(x_n, y_n) - f(x_{n-1}, y_{n-1})].$$

A suitable Adams–Moulton formula for use with this is
$$y_{n+1} = y_n + (h/2)[f(x_{n+1}, y_{n+1}) + f(x_n, y_n)]$$
corresponding to $q = 0$ in Eqn (6.55). The local truncation errors for these formulae are
$$\frac{5h^3}{12} f^{(2)}(\xi, y(\xi)) \quad (x_{n-1} < \xi < x_{n+1}),$$

and

$$-\frac{h^3}{12} f^{(2)}(\eta, y(\eta)) \quad (x_n < \eta < x_{n+1}),$$

respectively.

A suitable starting method for both the Adams–Bashforth and Adams–Moulton methods is the modified Euler method, given by Eqn (6.27).

# Bibliography

The following books could be consulted in conjunction with this text. Those marked with an asterisk (*) are more advanced than the others.

BECKETT, R., and HURT, J. *Numerical Calculations and Algorithms*, McGraw-Hill, London (1967).

BURKILL, J. C. *A First Course in Mathematical Analysis*, Cambridge University Press (1967).

CARNAHAN, B., LUTHER, H. A., and WILKES, J. O.* *Applied Numerical Methods*, Wiley, New York (1969).

CONTE, S. D. *Elementary Numerical Analysis*, McGraw-Hill, London (1965).

FORSYTHE, G. E., and MOLER, C. B.* *Computer Solution of Linear Algebraic Systems*, Prentice-Hall, Englewood Cliffs, N.J. (1967).

GROVE, W. E. *Brief Numerical Methods*, Prentice-Hall, Englewood Cliffs, N.J. (1966).

HENRICI, P.* *Elements of Numerical Analysis*, Wiley, London (1963).

HOHN, F. E.* *Elementary Matrix Algebra*, Collier-Macmillan, London (1964).

JENNINGS, W. *First Course in Numerical Methods*, Collier-Macmillan, London (1964).

MASSEY, H. S. W., and KESTELMAN, H., *Ancillary Mathematics*, Pitman, London (1959).

MOURSUND, D. G., and DURIS, C. S. *Elementary Theory and Application of Numerical Analysis*, McGraw-Hill, London (1967).

NATIONAL PHYSICAL LABORATORY* *Modern Computing Methods*, Notes on Applied Science No. 16, H.M.S.O., London (1962).

NOBLE, B. *Numerical Methods*, Vols 1 and 2, Oliver & Boyd, Edinburgh (1964).

RALSTON, A.* *A First Course in Numerical Analysis*, McGraw-Hill, New York (1965).

RALSTON, A., and WILF, H. S.* *Mathematical Methods for Digital Computers*, Vols 1 and 2, Wiley, New York (1966).

SCHEID, F., *Numerical Analysis*, Schaum's Outline Series, McGraw-Hill, New York (1968).

# Index

[*N.B. Words in bold type are to be repeated before succeeding items*]

Absolute error, 11
Algebraic equations, 83
Algorithm, 5
Asymptotic error constant, 98

Back substitution, 110
Backward difference operator, 37
Binomial function, 41

Chopping, 9
Constructive proof, 22
**Convergence** theorem, 6; linear, 95; geometric, 95; order of, 98
Corrector formula, 145

Decimal places, 9
Degree of precision, 61
Differentiation, 2
Direct methods, 108

Error bound, 1
**Euler's method**, 131; modified, 135
Existence theorem, 6
**Explicit single-step** method, 138; formula, 144
Exponent, 112

Forward difference operator, 33

Heun's method, 135

Ill-conditioning, 117
Implicit formula, 144

Integration, 3
**Interpolatory** points, 21; polynomial, 20
Interpolating quadrature, 53
Iterates, 89
Iteration function, 89
Iterative methods, 108

**Linear** algebraic equations, 4; interpolation, 24
**Lipschitz** condition, 84; constant, 85
Loss of significance, 15

Main diagonal, 108
Mantissa, 112
**Matrix** dense, 108; sparse, 108; band, 108; diagonal, 108; tri-diagonal, 108
Method of undetermined coefficients, 81
Multi-step formula, 144

Nested multiplication, 17
**Numerical analysis**, 5; defined, 7

**Order** of truncation error, 59; of convergence, 98; of iteration function, 98
Ordinary differential equations, 4

Partial pivoting, 112
Percentage error, 11
Pivots, 111

**Polynomial** approximation, 1; equation, 83
Predictor formula, 145

Quadratic convergence, 99
Quadratic formula, 65

Rectangle rule, 67
Relative error, 11
Residual vector, 117
Roots, 83
**Rounding error**, 8; local, 132

Significant figures, 9
Simpson's rule, 67

Single-step method, 140
Steffensen iteration, 97
Strict diagonal dominance, 120
Superlinear convergence, 97
Synthetic division, 18

Tabular points, 21
Transcendental equation, 83
Trapezium rule, 67
**Truncation error**, 7; cumulative, 131

Uniqueness theorem, 6

Zeros, 83